Machining Solutions

Machining Solutions

The Art & Science of Metal Cutting

URMAZE NATERWALLA

ISBN: 978-1-78324-141-5

Published by Wordzworth
www.wordzworth.com

This book is dedicated to my loving, sweet, funny, and talented daughter Sanaz Naterwalla.

CONTENTS

LIST OF TABLES

ABSTRACT

This book is intended to coach a reader through the very fundamentals of metal cutting and related best practices, all the way through some advanced machining solutions. Special attention is given towards keeping the content practical and focused towards modern day machining, by clear demonstration of the evolution of the machining concepts. A universal 'Litmus Test' that was applied towards the practicality of this book was by following two simple internal checks:

1. 'Can I use this content in an actual production environment?'
2. 'I wish the material was taught to me in that way.'

This book is intended to be a valuable recourse not only for beginners (such as students, and newly hired employees), but also for advanced users (such as graduate students, experienced engineers, research personnel, and instructors). Additionally, the content is particularly tailored for machine shop employees (operators, & maintenance), and technical employees (NC programmers, cutting tool designers / engineers). These readers would benefit from:

- In-depth understanding of machining concepts from their origins.
- The material could be directly implemented into everyday jobs almost immediately.
- Professional growth by way of effective & practical problem solving.
- Learning best practices that have been passed down over the generations.
- Lessons on optimally selecting machine parameters, as well as optimizing processes.

Emphasis needs to be given to the differentiations offered in this book. First, great care has been taken to research and explain the roots of machining concepts. Second, where applicable, the needed equations have been clearly derived with no gaps in their evolution. Third, the level of detail has been painstakingly filtered and organized based on the needs of the end user. And finally, consistent efforts have been made to link up the various concepts explained. This allows the user to mature from the very fundamental concepts of metal cutting, to being able to connect it with advanced machining solutions. The logical thinking patterns shown, will allow the end user to think on the spot in a stress filled production machining environment, and arrive at confident machining solutions.

ACKNOWLEDGMENTS

I would like to express my humble gratitude to my wife Jasmine, for her unrelenting love and support throughout my career, and in encouragement during the scripting of this text. I would like to dedicate this book to my daughter Sanaz, for being exactly who she is. She has been the silent inspiration, I never thought I would witness. I would like to thank my sweet grandmother Pillo, my mother Yasmin, and my father Dhunji for everything you have done for me.

I truly believe that education comes in all forms. Over the years, my learning has stretched far beyond than the mainstream academic offerings, to friends, mentors, and colleagues. My real schooling has come from countless conversation around production machines, over coffee talks, long drives and flights, sales & service calls, seminars, production failures, and working meals. Particular thanks to Mr. T.K Wong for teaching me the hands-on skills of manual machining. I am forever grateful to Dr. Ciro Rodgiues for his friendship, and mentoring me through my research beginnings at The Ohio State University. Sincerest thanks to Mr. Brian Hoefler for all the support, encouragement and leadership, early in my professional testing research career. Deepest thanks Mr. Robert Sullivan, his friendship, and for unselfishly sharing his wealth of knowledge through my early career. Special thanks to Dr. John Shih for teaching me the true meaning of leading by example, career guidance, and helping me navigate through the dynamics of large complex corporations.

This book would not exist were it not for the support of the wonderful folks at OMIC (Oregon Manufacturing & Innovation Center). This rarest of successful amalgamations where industry, Academia, and government, came together to yield a better manufacturing world has given this humble text a chance to help. OMIC is driven solely on a simple principle 'that if the water table goes up, all ships will rise'. Sincerest thanks to Mr. Craig Campbell in supporting me and this text, and to Mr. Grant Kirby for your guidance on its publishing.

Special thanks to senator Betsy Johnson for her leadership in supporting the resurgence of manufacturing in the state of Oregon, and recognizing the necessity of steering the youth towards the skills trades. Her vision has provided this book to serve as a platform for manufacturing excellence.

I am obnoxiously proud of our profession, and I offer this book to possibly help support its future evolution. I am forever grateful to the family of manufacturing and machining people throughout the world. You know who you are. I am thankful for all our experiences, conversations, arguments, challenges, and most of all your sense of humor. Speaking of which...

Finally, thank you to all the cutting tools and machine spindles that have sacrificed their lives in the line of duty, so that we could learn. You have paid the ultimate price for our human short comings, and materialistic greed. May you all rest in peace, in chip heaven.

1

INTRODUCTION TO MACHINING CONCEPTS

If we step back and think about it, machining is an elegant yet violent art form. It is simply the relatively controlled collision between the material to be cut, *'the workpiece'*, and the *'cutting tool'* used to cut it into the desired shape. As and when possible, we are simply trying to connect the science to this magnificent historic art form, so as to arrive at a more repeatable, reliable process for modern day production needs.

This relative movement between the workpiece and the cutting tool can broadly be categorized into:

- Milling
- Turning
- Drilling

1.1 MILLING

Visualize a lawnmower used to cut grass, a blender or food processor used for making smoothies, or a buffing wheel used to wax a car (***Figure 1***, & ***Figure 2***). In each of these everyday examples the respective cutting tools (i.e. the blades, & buffing wheels) are spinning at a significantly high speed, and are being maneuvered into or presented with a relatively slower travelling workpiece (i.e. the grass, food, car body). The understanding of these relative movements basically explains the differentiation of *'Milling'*. Some clear points of demarcation for Milling in metal cutting are:

THE CUTTING TOOL

Is revolving and provides the Speed (aka RPMs or revolutions per minute).
Is usually harder, tougher and more heat resistant than the workpiece.

THE WORKPIECE

Is often fed slowly into the Cutting Tool (aka FEED or inches per minute or mm per min).

NOTE: *Depending on the machine, the workpiece can also be stationary, and the machine FEEDs the fast rotating Cutting Tool into the workpiece.*

Figure 1. The Milling Concept.

Another unique and often missed differentiation in Milling is that by its own nature is an '***Intermittent Operation'***. This simply means that the cutting edge or blades go in and out of the cut. The small piece of cut material, commonly called a '***Chip'*** (or the grass, food particulates, and buffing compound in our everyday examples), is naturally broken when the cutting edge exits the workpiece. It is not long stringy and continuous as will be demonstrated in the explanations of Turning and Drilling.

Figure 2. The High Velocity Milling Concept.

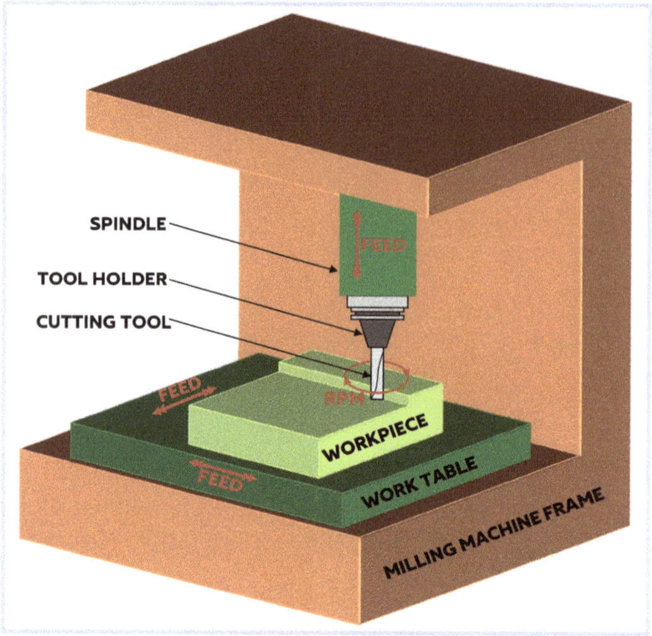

Figure 3. The Milling Setup.

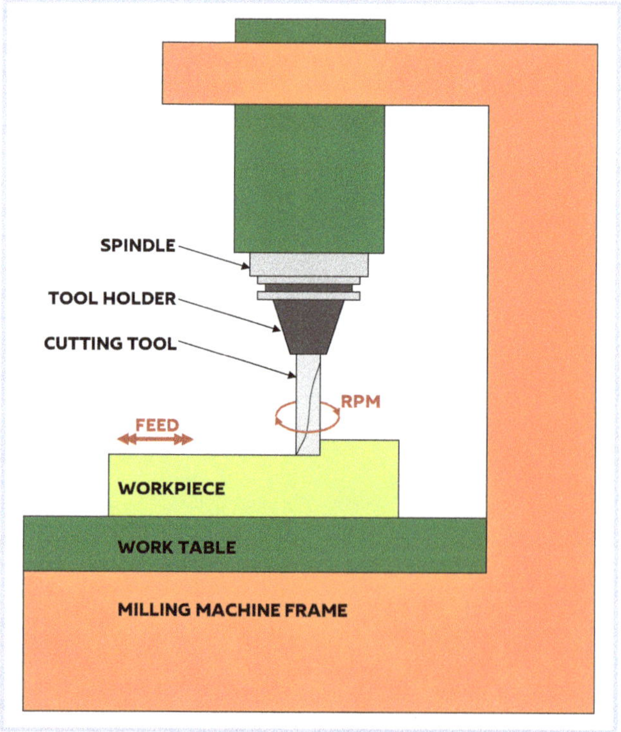

Figure 4. The Milling Setup (Side View).

1.2 TURNING

Visualize a manual pencil sharpener, or a fruit peeler. In each of these everyday examples the workpiece (i.e. the pencil, & fruit) are spinning at a significantly high speed, and are being maneuvered into or presented with a slower travelling cutting edge (i.e. the sharpener blade, or peeler knife edge). The understanding of these relative movements basically explains the differentiation of *'Turning.* The machine used in turning operations is commonly known as a Lathe. Some clear points of demarcation in Turning in Metal Cutting are:

THE CUTTING TOOL

Is stationary, or is fed relatively slowly into the workpiece (aka FEED).
Is usually harder, tougher and more heat resistant than the workpiece.

THE WORKPIECE

Is revolving and provides the Speed (aka RPMs or revolutions per minute).

NOTE: *The machine FEEDs the faster rotating workpiece into the cutting tool.*

Another unique and often missed differentiation in Turning is that by its own nature is a '***Non-Intermittent Operation'***. This simply means that the cutting edge or blades are continuously engaged in the cut. The cut material, commonly called a '***Chip'*** (or the pencil shaving, fruit peel, in our everyday examples), tends to be long and stringy, and breaks when the cutting is terminated.

Figure 5. The Turning Concept.

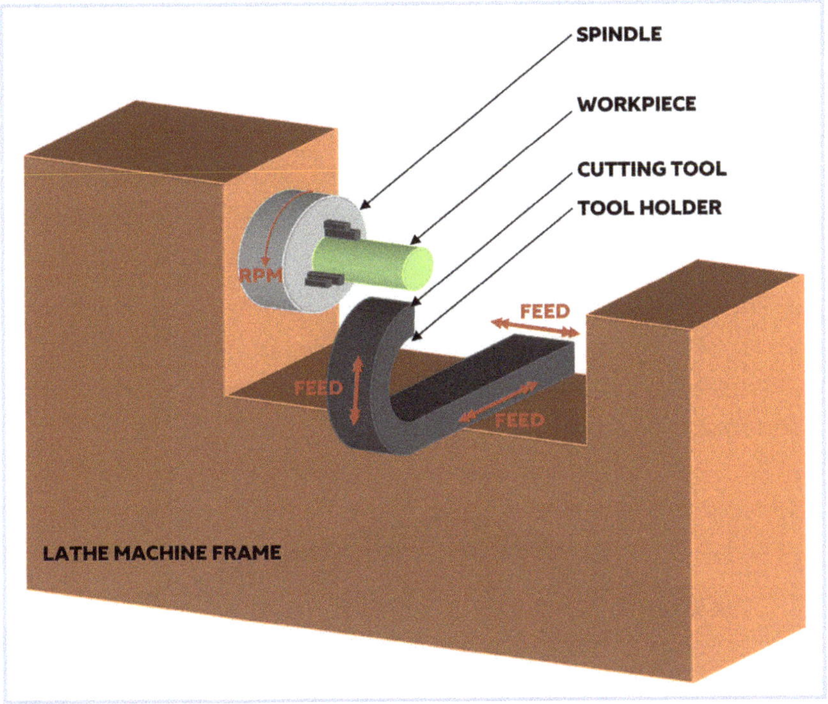

Figure 6. The Turning Lathe Setup.

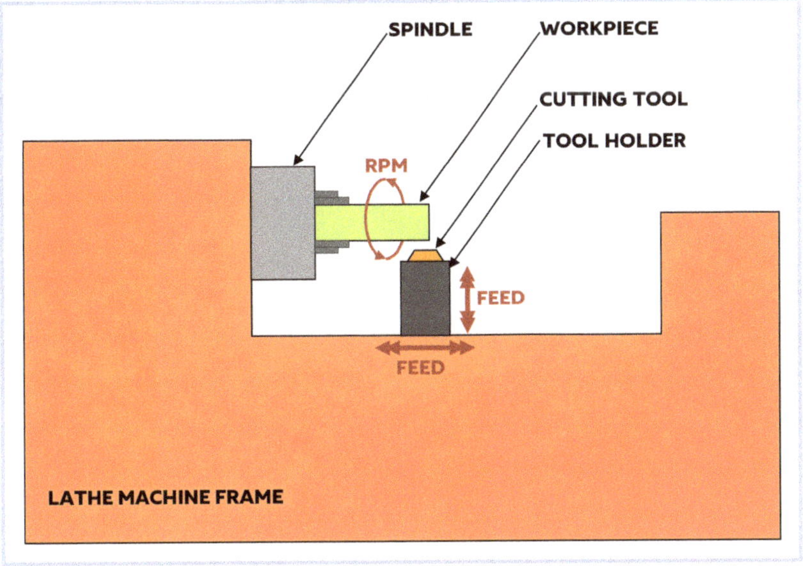

Figure 7. The Turning Lathe Setup (Side View).

1.3 DRILLING

Drilling is simply the metal cutting technique employed for hole making. This hole making can occur on milling machine, or a lathe, or a dedicated drill press. The point of differentiation in a drilling process over milling or turning is that drilling occurs in a singular axis of movement between the tool & workpiece (vis. Along the length of the hole being made).

In drilling, irrespective of whether the cutting tool provides the RPMs (Milling Machine or Drill Press), or the workpiece provides the RPMs (Lathe); the cutting edge of the drill bit is continuously in the cut. Therefore, by its nature drilling is a ***'Non-Intermittent'*** operation.

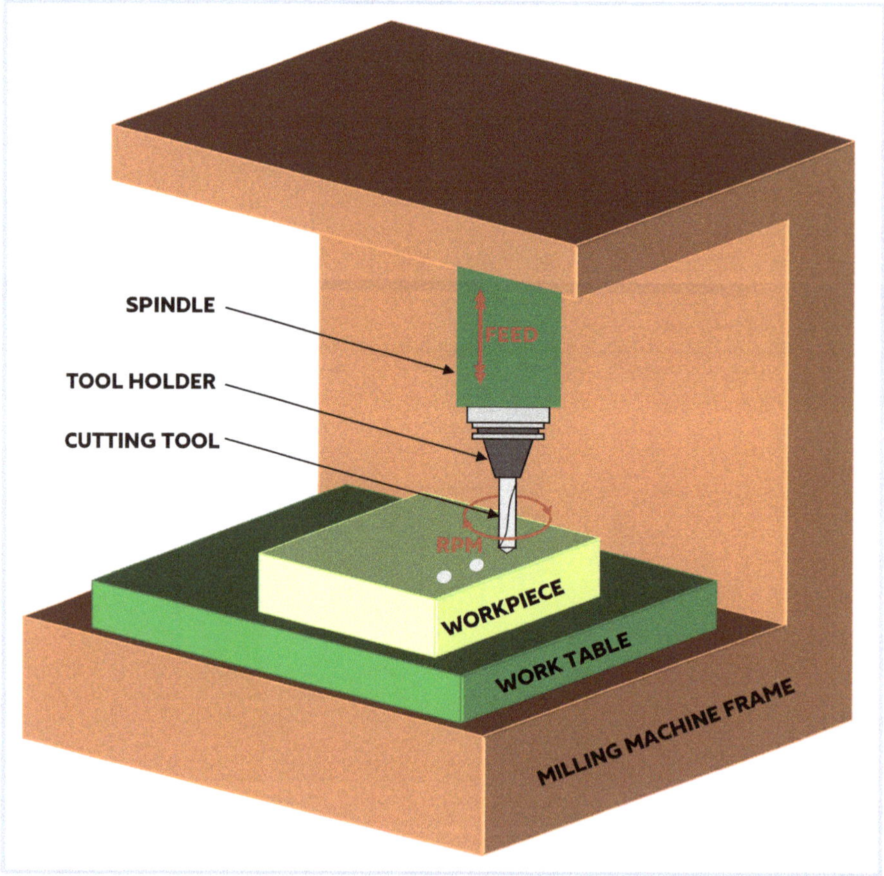

Figure 8. The Hole Making Machine Setup.

Another unique and often missed differentiation in Drilling is that the cutting tool (Drill Bit) is completely submerged in the hole being formed, and special attention needs to be given towards evacuating the chips (via grooves or flutes designed along the length of the drill bit). A drill bit pierces into the workpiece almost like a rotating arrow. Thus, it should be recognized the center point of the drill bit is always culminating into a single cutting edge. This differentiation will be recognized later during calculations of FEED during drilling.

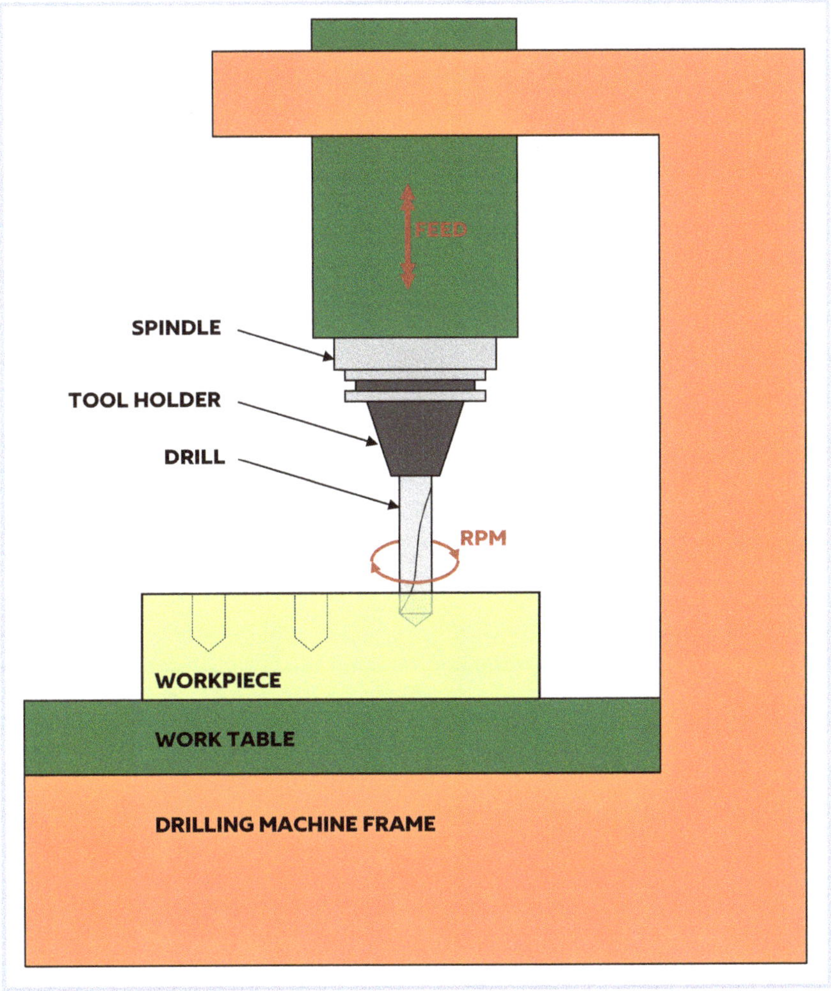

Figure 9. The Hole Making Machine Setup (Side View).

2

INTRODUCTION TO CUTTING TOOLS

Experience has shown that concepts of machining when taught and explained for milling usually carry over very well into turning and hole making. Therefore, in this text the metal cutting concepts will be initiated by way of milling, and then carried over into turning and hole making. A significant portion of this chapter is dedicated toward recognizing the commonly used nomenclature in metal cutting. In an industrial environment, this nomenclature often becomes abbreviated and slang like. These interpretations will be recognized throughout the text. Finally, there is a well thought out repetitiveness in exposing the reader to the appropriate level of detail on each of machining concepts described throughout the text. Certain fundamental nomenclatures and lingo need to be well engrained prior to deep diving the details of the machining. In other words, be patient, all the dots will eventually get connected.

2.1 SOLID CARBIDES & INDEXABLE CUTTING TOOLS

Milling cutters can broadly be separated into two distinct and necessary cutter families: Solid Carbides, and Indexable Cutting Tools.

2.1.1 SOLID CARBIDE CUTTING TOOLS

A Solid Carbide cutting tool is a more historical cutting tool that traces its roots to ancient blacksmiths. Simply put, the cutting edge and other necessary geometry is ground onto a single solid piece of carbide to create the required cutting tool (***Figure 10***). Solid Carbide Cutting Tools started out historically as cutter made out of High Speed Steels (HSS). Later on, as technology evolved, the industry paid a heighted awareness to the Cobalt (Co) content of these HSS tools. These cutters were labelled or nicknamed as Cobalt Cutters. As time progressed, greater attention was given not only to the Co content, but also to the Tungsten Carbide (WC) alloying element. This is when these cutters were nicknamed Carbide cutter. Modern day Carbide Cutters not only have other alloying elements besides Tungsten Carbide (WC) such as Nickel (Ni) or Chromium (Cr), but are also coated with heat resistant coatings such as Titanium Nitride (TiN), or Aluminum Oxide (Al_2O_3). These coated carbide tools are also labelled as ***'Carbide Tools'*** in modern day industry.

2.1.2 INDEXABLE CUTTING TOOLS

As manufacturing evolved, the metal cutting industry recognized the need for substantially larger sized cutting tools than those that were commonly available as Solid Carbides. Additionally, it would be extremely cost prohibitive to continue to build larger and larger Solid Carbide cutting tools. In an everyday example: a garden shovel is made of a wooden handle, and a steel shovel head. It would not be very efficient (cost & function) to build the entire shovel out of steel. Analogous to this shovel example, the metal cutting industry engineered the Indexable Cutting Tool. Here pieces of Carbide (relatively more expensive & less available) were assembled by way of fasteners onto a cutter body which was predominantly made out of steel (relatively cheaper & more available). These pieces of Carbide are also commonly called INSERTS. These inserts when engineered smartly and placed strategically onto a cutter body, can create a continuous cutting edge (***Figure 10***).

As you can imagine, these Inserts could be produced in various geometrical configurations, the most popular of which were rectangular cubes, buttons, and triangles. Such three-dimensional insert configurations could then be ground with multiple functional cutting edges. For example, a rectangular cube could have at least four fully functional cutting edges. Consequently, after the first cutting edge got worn, the user could then rotate or 'INDEX' the insert to present the next unused cutting edge. This yielded the maximized use of the piece of

Carbide insert, and cost efficiency. Hence, these Carbide Inserted cutting tools got further nicknamed ***'INDEXABLES'*** in modern day industry.

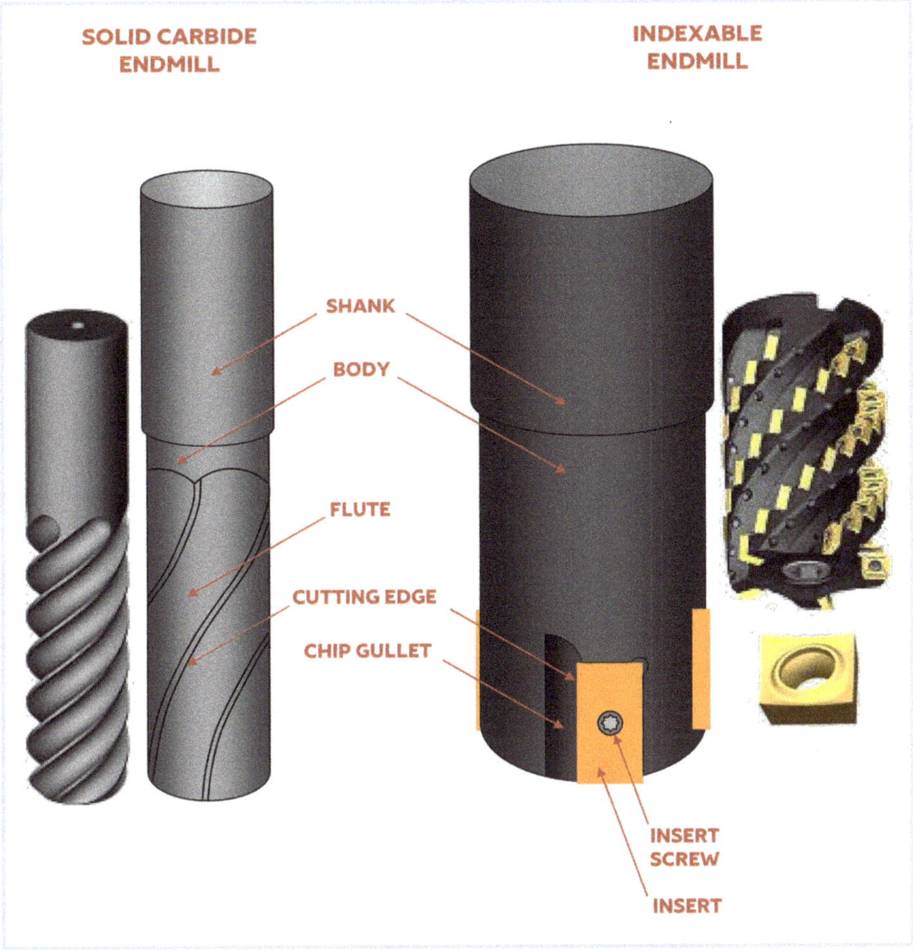

Figure 10. Solid Carbide vs Indexable Endmills.

2.2 RAKE AND FLANK FACES

Throughout this text much of the learning comes in the familiarization of metal cutting nomenclature. Before we go much further it is best to talk about rake and flank faces of a cutting tool. At the most basic level, the rake face surface meets up with the flank face surface to create the cutting edge. The cutting edge starts the formation of a chip. The chip then curls up on the rake face. As seen in ***Figure 11***, the rake face can singularly be distinguished as the face on which the chip curls. Analogous to a garden rake which rakes up leaves on its leading face, the cutting tool's rake face simply rakes up the chip. We want the cutting tool to create the chip, and yet not rub along the surface of workpiece. The flank faces are basically the faces that are adjacent to the rake face, as seen in ***Figure 11***. One of the primary purposes of the flank face is to provide clearances from the surface of the workpiece so as to avoid any rubbing of the cutting tool.

Although a bit premature, it may be relevant at this point to quickly recognize the rule of thumb in regards measuring tool wear on each of these faces (which will be discussed in a later chapter). During a tool life test, the tool wear is always measured on the flank face. Although it is important to observe the wear on the rake face, taking wear measurements on it can be optional.

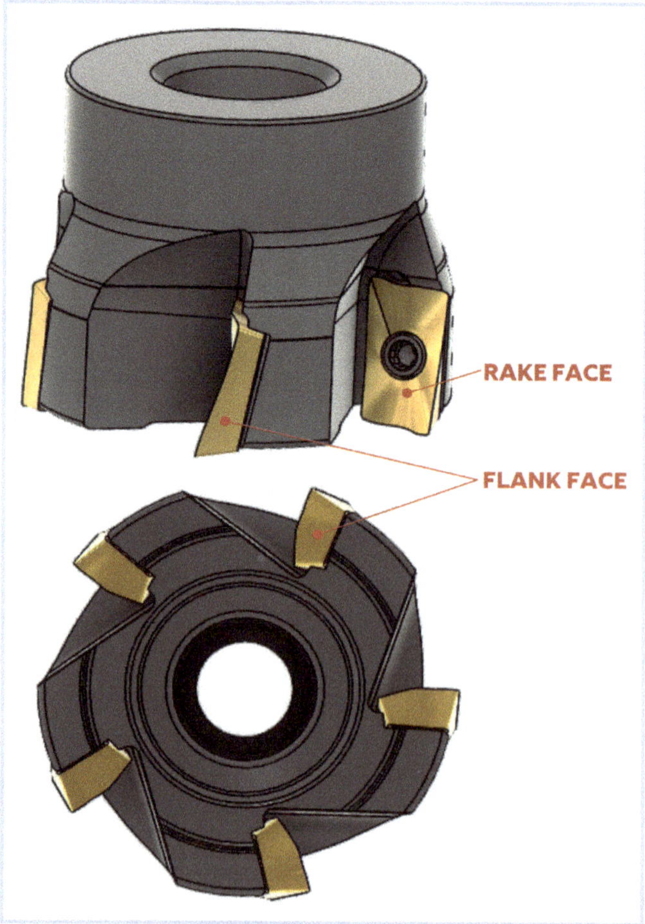

Figure 11. Rake & Flank Faces.

2.3 ORTHOGANAL AND OBLIQUE VIEWS

As we read some more of the formal metal cutting publications, the words get fancier for no apparent reason. Particularly amongst the folk who work in academia and research, who do Finite Element Analysis and other modeling of cutting tools, us terms such as orthogonal and oblique views. As seen in ***Figure 12***, an orthogonal view is simply a 2D representation of a cutting tool, and an oblique view is a 3D rendition of the same.

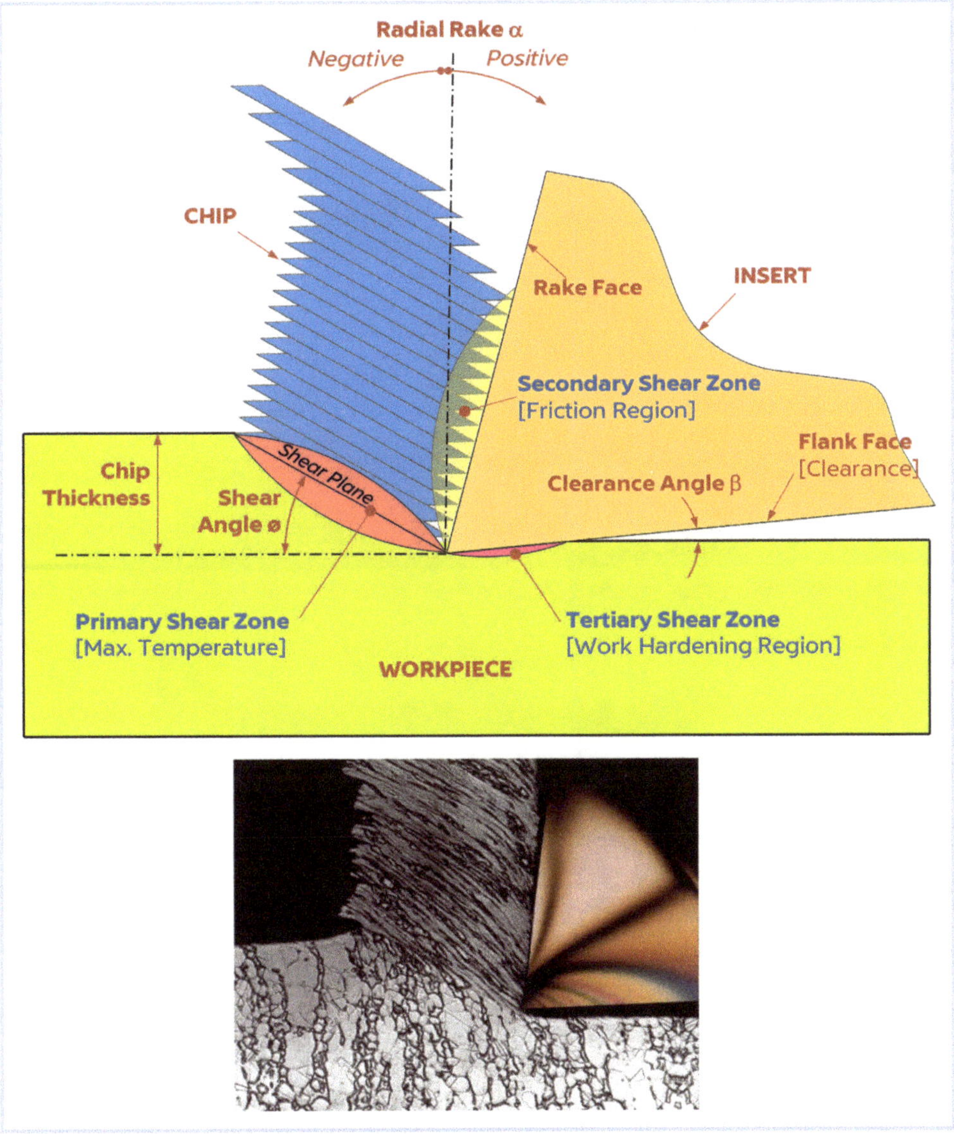

Figure 12. Orthogonal View in Machining.

2.4 A CUTTER ASSEMBLY

At the risk of sounding redundant a cutter assembly is referred to as the assembly of the cutting tool (with inserts or solid carbide) to the holder (***Figure 13***). This task although sounds simplistic, is quite important to get correct in a manufacturing environment. Tool assemblers pay particular attention to a critical dimension known as the Gage Length, and Clear Length. Deviation from this simple attention to detail has been the root cause of many a machine crashes, scrapped parts, and catastrophic failures. There are instances where the cutting tool body and holder are made as one piece, and do not need to be assembled. These are referred to in the industry as Integral Cutters.

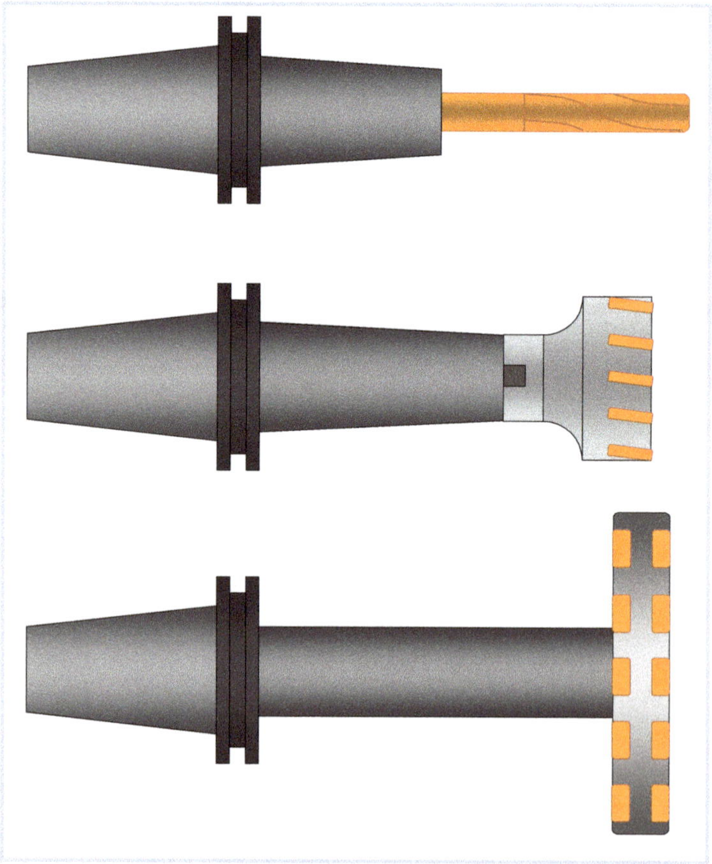

Figure 13. Cutting Tool Assemblies.

2.5 THE BASIC DNA OF AN INSERT

As seen in ***Figure 14***, an insert is fundamentally comprised of a substrate, and a coating. The primary purpose of the substrate is to provide the hardness and toughness characteristics of the cutting tool (discussed in detail later), while the coating provides the thermal resistance along with some surface lubricity to the cutting tool. Of course, the coating has some hardness and toughness characteristics, just as the substrate inherently has some capability to fight heat. However, the former statement is a good way to universally understand the performance characteristics as well as trouble shoot failures of the cutting tool when they occur. This basic rule of thumb carries over from inserts into a modern day coated solid carbide tool as well.

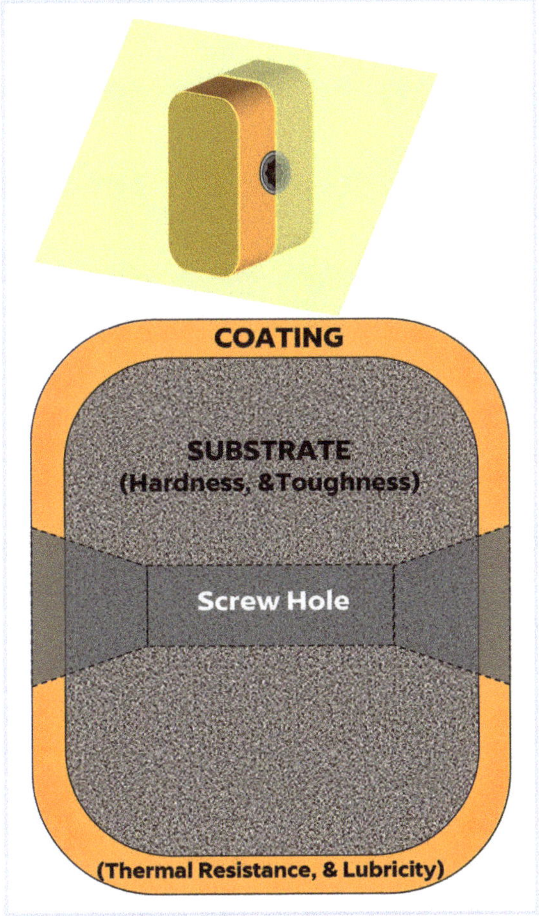

Figure 14. The Cross Section of an Insert.

3

HARDNESS & TOUGHNESS

From here on pretty much every subsequent concept discussed in this text can be related to the fundamental material mechanical properties known as hardness and toughness. Clarity on these properties can prove to be a very powerful tool in not only troubleshooting machining operations, but also optimizing them. Although intuitive to some, this nomenclature is often thrown around inappropriately by people in the trade.

3.1 HARDNESS

Simply put, hardness is the ability of a material to resist abrasion. Harder materials tend to be brittle, and don't have much ability to bounce back when put under tension or bending loads. As seen in ***Figure 15***, a diamond, a brick, a cinder block, or a glass vase are perceived materials that are hard. Additionally, a piece of sand paper can be looked as being hard. Its primary job of sandpaper to resist abrasion while eroding layers of material from the surface of the workpiece on which it is rubbed.

3.2 TOUGHNESS

Toughness is the ability of a material to resist impact. Tough materials tend to be malleable, and can bounce back when put under tension or bending loads. A visualizing technique (***Figure 15***): A football player or boxer is often labeled as tough. In other words, the player hopefully has the ability to bounce back up after being subject to impact or blows. Incidentally a sponge or rubber band are inherently very tough (but not hard). They can both bounce back when subjected to deformation.

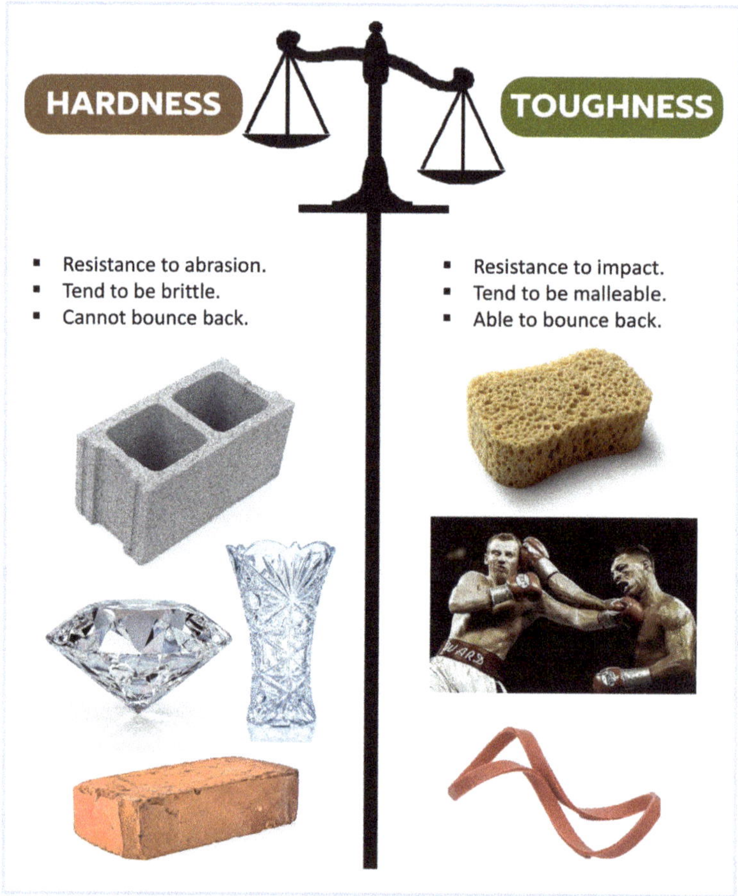

Figure 15. Hardness & Toughness.

Hardness and toughness are often on each side of the balance scale of material properties. If a material is hard, it is consequently less tough and vice versa. Just as a diamond is extremely hard, it is not very tough at all. A sponge although very tough, is not very hard at all. A healthy cutting tool is a challenging balance of hardness and toughness (along with thermal resistance) to best suit cutting the intended workpiece.

There are various real-life examples to materials and objects that are a harmonious blend of hardness and toughness to yield peak functionality. Let's look at a car tire. Is it hard or tough? Actually, it a combination of both (***Figure 16***). Think of all the bumps and impacts that a car tire would have to overcome over its useful life. Therefore, it definitely needs to have toughness characteristics. Additionally, a car basically propels forward on the merits of friction between the road and the tires. Therefore, it also needs high resistance to abrasion, or it would simply disintegrate within moments of use.

Figure 16. Hardness & Toughness – The Car Tire Analogy.

3.3 HARDNESS & TOUGHNESS – THE ROAD ANALOGY TO CARBIDE

We find that the composition of a typical black top road is very analogous in composition and behavior to the substrate of a carbide cutting tool. This visualizing technique can greatly help the user to troubleshoot and optimize machining operations.

Although road building is a marvelous science in itself, for the sake of our discussion we will take a simplistic approach. A road as seen in ***Figure 17***, can be viewed as comprising of small rocks or gravel that are held together by a binder known as tar. The rocks in the road provide the hardness characteristics, while the tar binder provides the toughness characteristics of the road. This combination of rocks and tar make the road a blend of the required hardness and toughness.

If you split open a carbide cutting tool and look at it under a microscope (***Figure 18***), it looks very similar to the road analogy discussed earlier. The cutting tool substrate is comprised of tungsten carbide (WC) which is held together by a binder which is cobalt (Co). The Tungsten Carbide is analogous to the rock on the road, and provides the hardness characteristics of the cutting tool. Similarly, the Cobalt binder is analogous to the tar on the road and provides the toughness characteristics of the cutting tool.

Now a road can be purpose built for its function (***Figure 17***): a city highway, a race track, or an airport runway. As you can imagine an airplane runway could be subjected to some serious impact when an airplane lands on it, followed by abrasions as the large airplane mass comes to a stop from high speed. Relatively speaking, a racecar track would see significantly less impact but higher abrasion from the friction caused by the race tires spinning at very high rpms. The hardness and toughness properties of the required road are adjusted by numerous ways:

- The amount of rocks in the road.
- The amount of tar in the road.
- The size of the rocks in the road.
- The texture of the rocks in the road.
- Multi layering, and combinations of rock sizes.

The hardness and toughness characteristics of a cutting tool can also be adjusted similarly. Increasing or decreasing the Tungsten Carbide (WC) content or grains size, increases or decreases the hardness of the cutting tool respectively. Similarly, increasing or decreasing the Cobalt (Co) content or grain size, increases or decreases the toughness of the cutting tool respectively. It should always be remembered that in a system if hardness goes up, then toughness by default goes down, and vice versa.

In the machining world a common lingo to describe the grade of Carbide is the % Cobalt content. For instance, in most hard metals rough milling operations the popular Carbide grade of choice is the 10 to 12% Cobalt content. As seen in ***Figure 19***, the Cobalt can be visualized as the darker background of binder (or glue-like material) that holds together the Carbide 'rocks'.

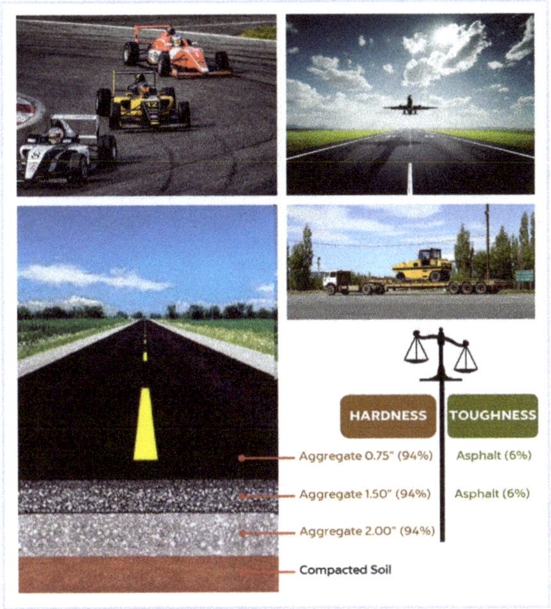

Figure 17. Hardness & Toughness in Carbide – The Road Analogy.

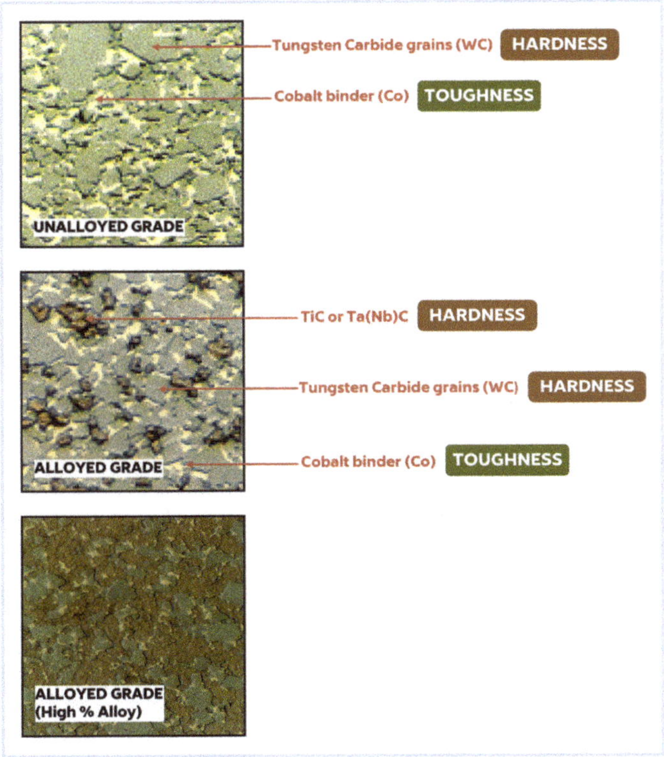

Figure 18. Tungsten Carbide (WC) Substrate.

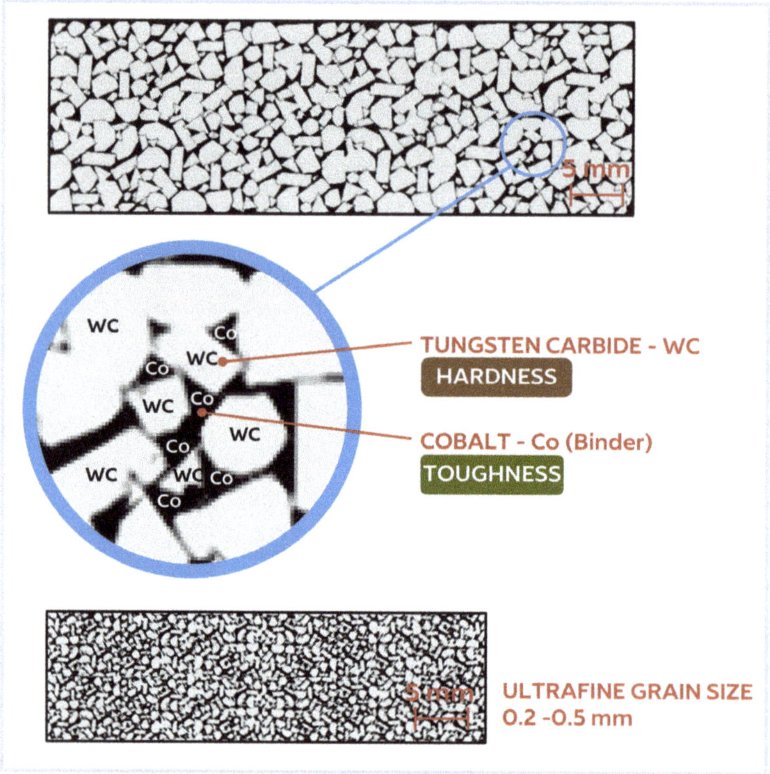

Figure 19. Visualizing Tungsten Carbide Substrate Matrix Composition.

3.4 ALTERING THE VARIABLES IN A CARBIDE TOOL SUBSTRATE

Let's march through some of the common scenarios as shown in ***Figure 20*** on the impact of the variables in a Carbide tool substrate on hardness and toughness:

- **As Co content increases:** The hardness decreases, & the toughness increases.
- **As WC content increases:** The hardness increases, & the toughness decreases.
- **As Co grain size increases:** The hardness decreases, & the toughness increases.
- **As WC grain size increases:** The hardness increases, & the toughness decreases.

There are some additional unique considerations to keep in mind on the above variable altering scenarios that can impact the cutter performance:

As Co grain size increases & wears: You can visualize the small binding ball-bearing are now larger marbles. This can now create larger gaps in between the bonds, as well have lesser bonding area of contact (***Figure 20***). Now as these larger Co grain sizes wear, they can simply be viewed as creating larger voids in the overall toughness of the carbide substrate.

As WC grain size Increases & wears: This will need further adjustments to the Co content and size to form good bonds. Additionally, as the larger WC grain size wears in a cut, it can break off and leave a relatively large void in the overall hardness of the carbide substrate.

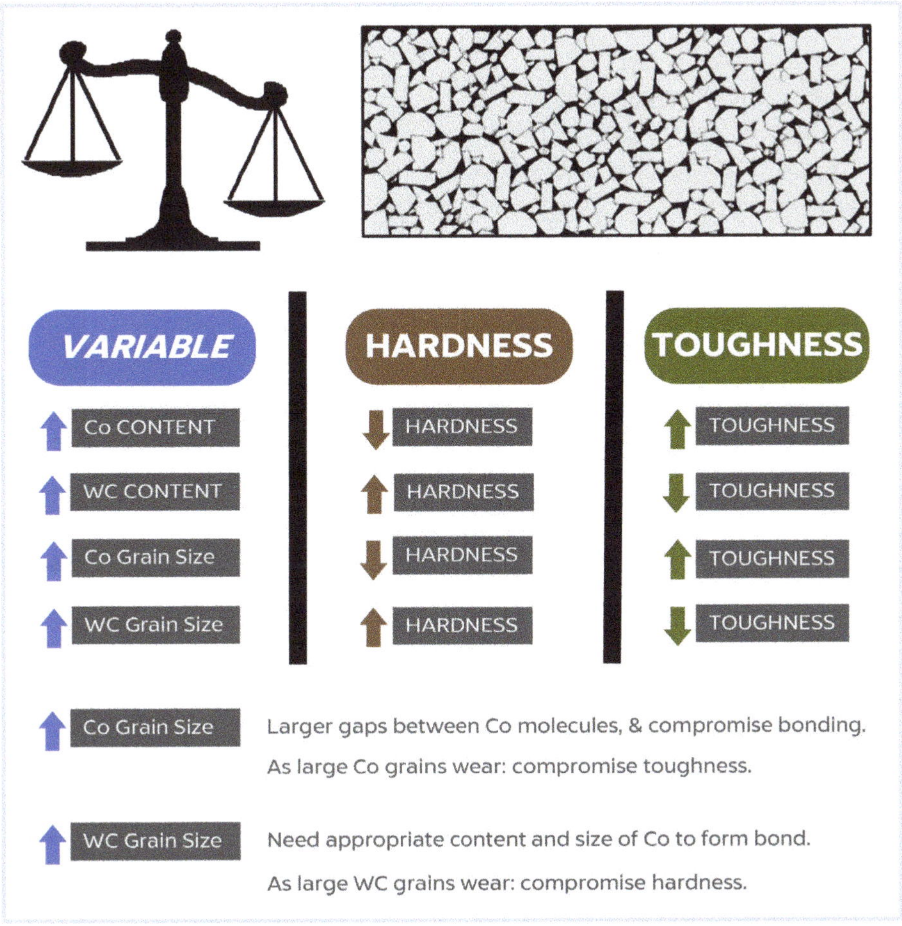

Figure 20. Altering the Variables in Carbide Substrate.

3.4.1 INDUSTRY APPLICATIONS BASED ON COBALT CONTENT & GRAIN SIZE

Cutting tool manufacturers are constantly adjusting the Tungsten Carbide substrate recipe to ideally suit the end-user's applications in various workpiece materials. With the intent of delivering repeatable-reliable performance over a tool-life that provides value for the customer's investment, the hardness & toughness is what is fundamentally adjusted. ***Figure 21*** shows a chart of typical combinations of Cobalt content and Tungsten Carbide grain size that yield substrates commonly used in various machining trades.

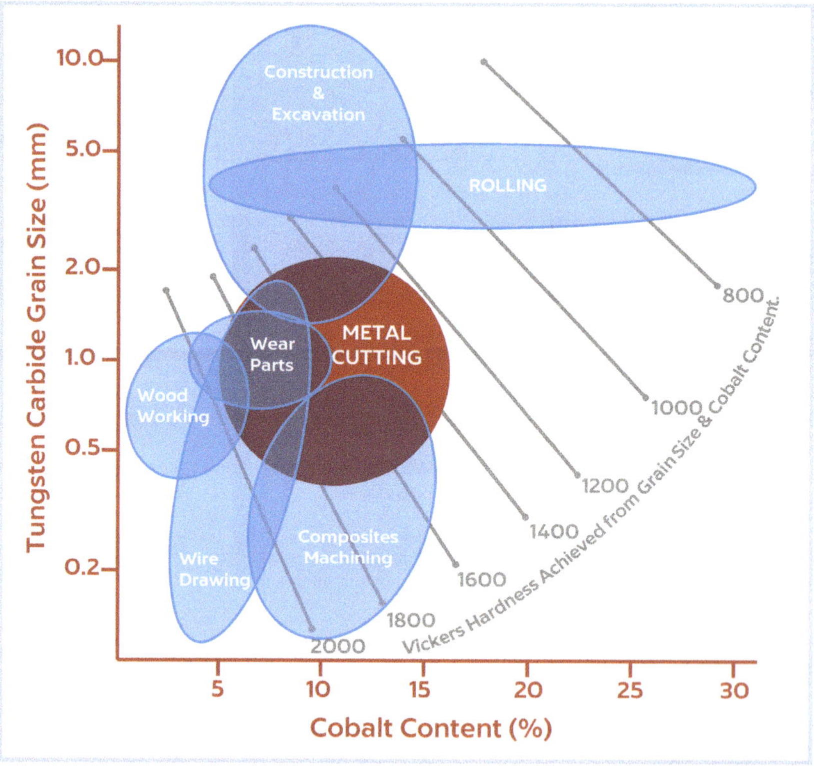

Figure 21. Typical Tungsten Carbide Grain Size & Cobalt Content Combination for Industry Applications.

4

MANUFACTURING CUTTING TOOLS

Teaching the details of cutting tool manufacturing has never been the intent of this text. The primary focus of this chapter is to provide the user with a high-level appreciation of the amount of care & attention that goes into manufacturing a cutting tool. It is by no means intended to completely breakdowns the various steps it takes to professionally manufacture a cutting tool. The level of detail covered is simply sufficient so as to help the user connect the dots to the application of cutting tools. Additionally, the text identifies how specific cutter manufacturing processes can become limiting to where the cutter can and cannot be used. Simply put, this chapter is a quick and relevant overview.

4.1 CUTTING TOOL SUBSTRATE MATERIAL FAMILIES

A high-level compartmenting of cutting tool material families is broken down as following:

Primary Substrate Materials

- Tungsten Carbides
- Cermets

Other Substrate Materials

- Ceramics
- SiNi, Al2O3

Substrate Tips

- CBN, Diamond

4.1.1 COMPOSITION OF TUNGSTEN CARBIDE SUBSTRATE MATERIAL FAMILY

The breakdown of the Tungsten Carbide substrate material family is as follows:

Primary Substrate Matrix Materials

- Titanium Carbide (TiC)
- Tantalum Carbide (TaC)

Substrate Binder Materials

- Nickel (Ni)
- Cobalt (Co)

Substrate Alloys, Matrix Material

- Tungsten Carbide (WC)
- Molybdenum Carbide (MoC)
- Titanium Carbonitride (TiCN)

4.1.2 COMPOSITION OF CERMET SUBSTRATE MATERIAL FAMILY

The breakdown of the Cermet substrate material family is as follows:

Primary Matrix Materials

- Titanium Carbide (TiC)
- Tantalum Carbide (TaC)

Binder Materials

- Nickel (Ni)
- Cobalt (Co)

Alloys, Matrix Material

- Tungsten Carbide (WC)
- Molybdenum Carbide (MoC)
- Titanium Carbonitride (TiCN)

4.2 CUTTING TOOL COATINGS

Some commonly used cutting tool coating materials are as follows:

- Titanium Nitride (TiN)
- Titanium CarboNitride (TiCN)
- Titanium Carbide (TiC)
- Titanium Aluminum Nitride (TiAlN)
- Aluminum Oxide (Al_2O_3)

Each of these coating come with their inherent advantages & disadvantages. Additionally, the coatings have specific application processes. These are summarized by the table in ***Table 1***. Details on PVD, and CVD coating process will be covered in a later section on carbide manufacturing.

COATING	ADVANTAGES	DISADVANTAGES	PROCESS
TiN	Best adhesion	Lowest wear resistance	PVD, MTCVD, CVD.
TiCN	Good adhesion	Fair wear resistance	PVD, MTCVD, CVD.
TiC	High wear resistance	Poor adhesion	PVD, MTCVD, CVD.
TiAlN	Mimics oxide with sharp edges	Limited thickness	PVD
Al_2O_3	Highest wear resistance. Chemically inert at high temps.	No sharp edges. Poor adhesion.	MTCVD, CVD

PVD – Physical vapor deposition

MTCVD – Moderate temperature chemical vapor deposition

CVD – High temperature chemical vapor deposition

Table 1. Coatings – Advantages, Disadvantages, and Application Processes.

4.3 CARBIDE MANUFACTURING

Figure 22 shows at a quick glance the various stages of carbide substrate manufacturing, starting from the ore of the raw materials all the way to the finished product.

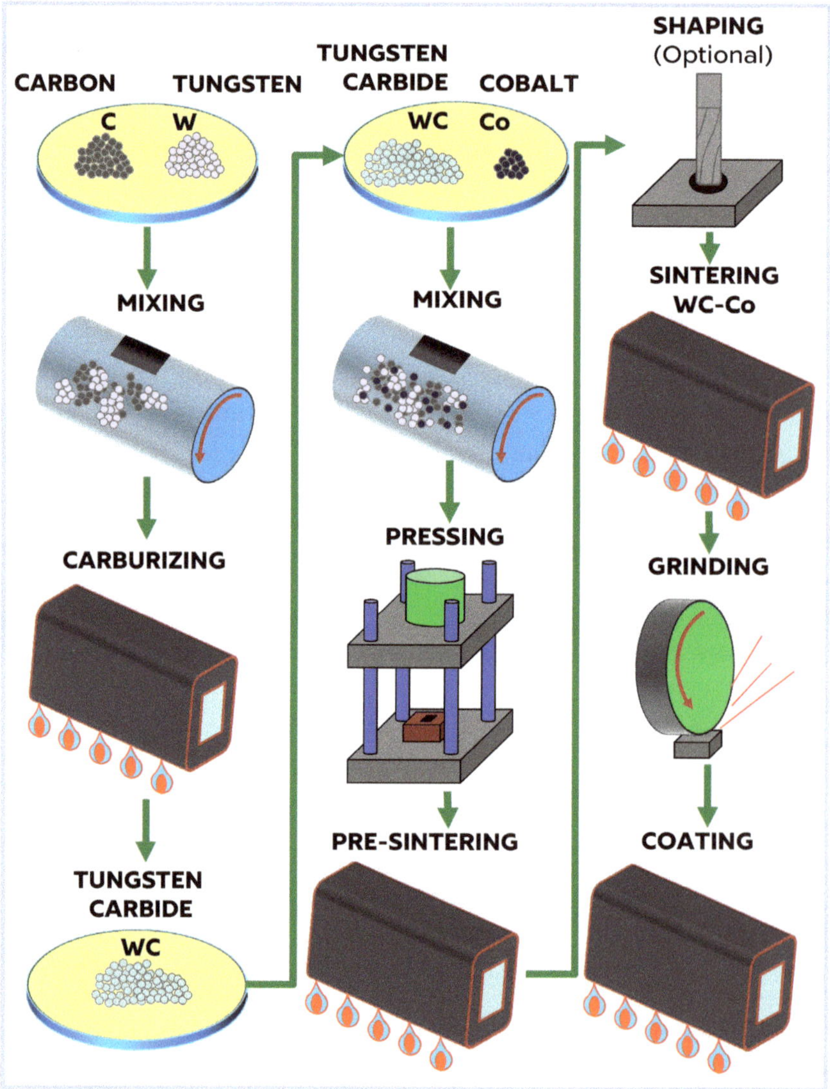

Figure 22. Carbide Tool Manufacturing – Substrate Manufacturing.

4.3.1 THE RAW MATERIALS

Taking it all the way back to fundamentals, Wolframite and Scheelite are crushed milled and leached with NaOH. The chemical reaction over the span of week creates Ammonium Paratungstate. It should be recognized that it is this Ammonium Paratungstate powder that determines the final Grain Size of the substrate. Heat drives out the Ammonia and water to get WO_3, which is then heated in dry Hydrogen gas to create Tungsten (W). The Tungsten is then mixed (milled) with Carbon (C) and then heated (Carburizing) to create Tungsten Carbide (WC), ***Figure 22***. Typical ratios of Tungsten to Carbide are 94 : 6 respectively. Similarly, in about the same ratio, other substrate options include Tantalum Carbide, and Titanium Carbide.

4.3.2 SHAPING THE CUTTING TOOL - GREEN STAGE

The Tungsten Carbide (or Tantalum Carbide or Titanium Carbide) is then combined with Cobalt (Co). These are then mixed (milled) while adding wax to the recipe. The wax gives the powder strength. This mixing process can sometimes take as long as a week. It is then compacted in a pressing process (7,250 to 21,750 PSI) to form a general shape of the cutting tool (only scaled up from the desired final product). At this point the cutting tool has a chalk like consistency, and is known as the 'Green Stage'.

4.3.3 CREATING CARBIDE BLANK FROM THE GREEN STAGE

The Carbide blank in the Green Stage is now sintered in a furnace (2550° F to 2900° F). This allows the wax to burn off. This process is also referred to as sinter-hipping (Hot Isostatic Pressing), is sintering under uniform pressure, which allows for near zero porosity. As seen in ***Figure 23***, the sintering process causes shrinkage of the blank. On average the linear shrinkage can range from 18% to 26%, while the volumetric shrinkage can range from 45% to 60%. It is found that the amount of shrinkage is related to the Cobalt content. When the wax burns off during sintering, it is the Cobalt the moves in to fill in the pores between the Tungsten Carbide grains. Tolerances from shrinkage can be hard to maintain. Designers use specialized software to accurately predict the amount of shrinkage for a blank design.

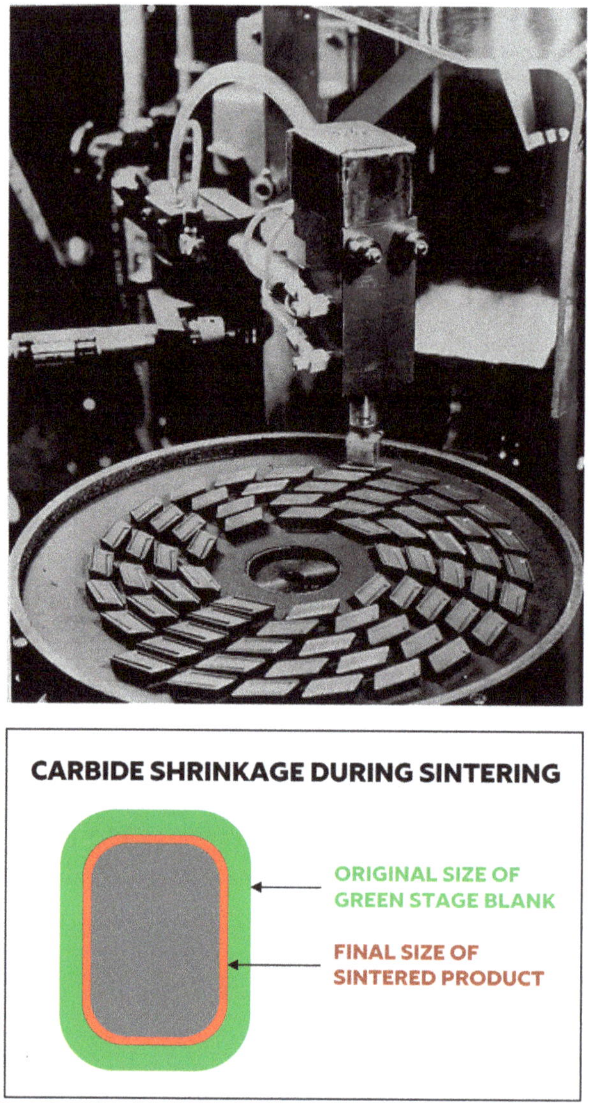

Figure 23. Carbide Tool Manufacturing - Substrate Sintering & Shrinkage.

Finally, if needed the insert blank is ground down to its final dimensions, ***Figure 24***. Since Carbide is very hard, the grinding is done with diamond impregnated wheels. Alternately, insert blanks that are left un-ground are called as-pressed. Each process come with their pros and cons.

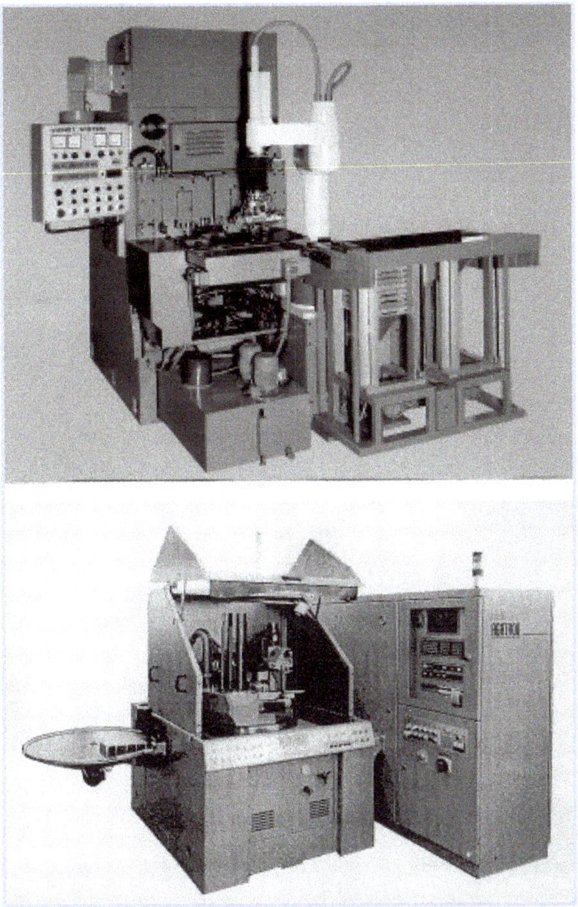

Figure 24. Carbide Tool Manufacturing – Substrate Grinding.

4.3.3.1 AS PRESSED BLANK

An as-pressed substrate blank tends to be more economical, and carry no residual stresses that are otherwise caused by the grinding process. Tolerances of 0.001" to 0.003" are possible in an as-pressed blank. It should be recognized that perfectly sharp edges are not possible in an as-pressed blank. Finally, the likelihood of leaching is least possible in an as-pressed blank.

4.3.3.2 GROUND BLANK

A ground blank tends to be more expensive as compare to an as-pressed one. It creates a more consistent edge condition and position. Tolerances of +/- 0.0005" can be maintained on a ground blank. By way of the grinding process, it is possible to create sharp edges. The grinding of a blank eliminates problems caused by Cobalt capping.

4.3.4 EDGE PREPARATIONS

The edge preparation is basically a dulling type treatment given a sharp edge. Simply put it is the same difference we see between a steak knife (sharp), and a butter knife (dull). Typical edge preparations include honing (rounding), Lands (chamfering), or a combination of hone and land.

Honing reduces chipping of a sharp cutting edge. Hones have also been known to reduce chatter. It is done by way of brushes, or by moving the tool through a slurry of specialized media. With these processes, hone tolerances of +/- 0.0005" can be achieved consistently.

4.3.5 CARBIDE SUBSTRATE CLEANING

Any insert that is to be coated must be cleaned because Contamination beneath coatings will cause poor adhesion of the coat to the substrate. Secondly, contaminants allowed into coating furnaces may cause damage to furnace mechanism (esp. In PVD vacuum type furnaces).

Inserts must pass "White Glove" type test to be considered clean enough.

4.3.6 CUTTING TOOL COATING PROCESSES

Cutting tool coating processes are differentiated by not only material of the coating, but also the eventual application of the cutting tool. These coating variables were touched on earlier in ***Table 1***. The common coating processes are:

- Physical Vapor Deposition (PVD).
- Chemical Vapor Deposition (CVD).
- Moderate Temperature Chemical Vapor Deposition (MTCVD).

4.3.6.1 PHYSICAL VAPOR DEPOSITION (PVD)

A high-level explanation of the Physical Vapor Deposition (PVD) process can be seen in ***Figure 25***.

More important, and aligned with the mission of this text is the understanding of the limitations and correct application of the PVD coatings:

PVD Process Highlights

- PVD is a lower temperature process (300-550°C).
- PVD is an electrically & magnetically driven reactive gas process.

- PVD is also a very low pressure (10-3mbar) process.
- PVD is a "Line of Sight Process".
- Typical PVD coating thickness ranges from 2-6 microns.

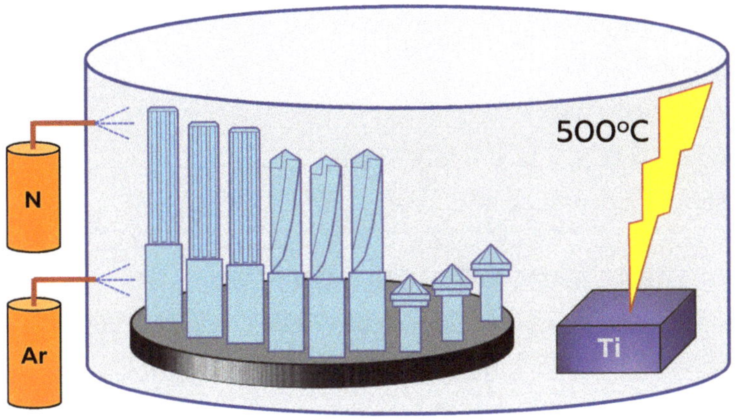

Electron beam ionizes metallic element (Ti or Al) which reacts with N or C on insert surfaces

GASSES USED IN PVD REACTION

$$2Ti + N_2 \rightarrow 2TiN$$

$$2Ti + N_2 + C_2H_4 \rightarrow 2TiCN + H_2$$

$$2Ti + 2Al + N_2 \rightarrow 2TiAlN$$

[NOTE: Other reactions following this logic are used to produce ZrN, CrN, etc.]

Figure 25. Physical Vapor Deposition PVD – Application Process.

PVD Coating Materials

- TiN, TiC, TiCN, TiAlN.
- Also, CrC, CrN, ZrN.
- [NOTE: No Al_2O_3 available yet.]

PVD Limitations

- PVD is a "Line of Sight Process". Similar to painting with a spray can, only visible features coated.
- Compressive stresses limits PVD to low coating thickness.
- Comparatively, the adhesion is marginal in PVD process.
- No aluminum oxide compositions currently available.

PVD Advantages & Usage

- PVD does not reduce the fracture toughness of the carbide.
- PVD helps protect the carbide substrate from heat and chemical attacks.
- The lower temperatures allow for coating sharp edges.
- Sharp cutting edge is required to avoid:
  - Breakout in cast iron.
  - Built up edge in aluminum.
  - Part movement due to unstable setup.
  - Excessive tool pressure and heat.

4.3.6.2 CHEMICAL VAPOR DEPOSITION (CVD)

A high-level explanation of the Chemical Vapor Deposition (CVD) process can be seen in ***Figure 26***.

More important, and aligned with the mission of this text is the understand of the limitations and correct application of the CVD coatings:

CVD Process Highlights

- CVD is a higher temperature process (900 – 1100°C).
- CVD is a temperature driven reactive gas process.
- The higher temperatures would destroy a sharp edge during coating. Therefore, the CVD requires edge honing.
- CVD is a higher pressure (50-500 mbar).
- In CVD the application determines coating thickness. Typically, CVD coating thickness ranges from 5-20 microns.

CVD Coating Materials

- TiN, TiCN, and TiC.
- HfN.
- Al_2O_3.

CVD Limitations

- CVD reduces the fracture toughness of the carbide.
- CVD lowers the substrate strength.
- Not likely used when:
  - Breakout is a concern.
  - Very fine finish or tolerance are required.

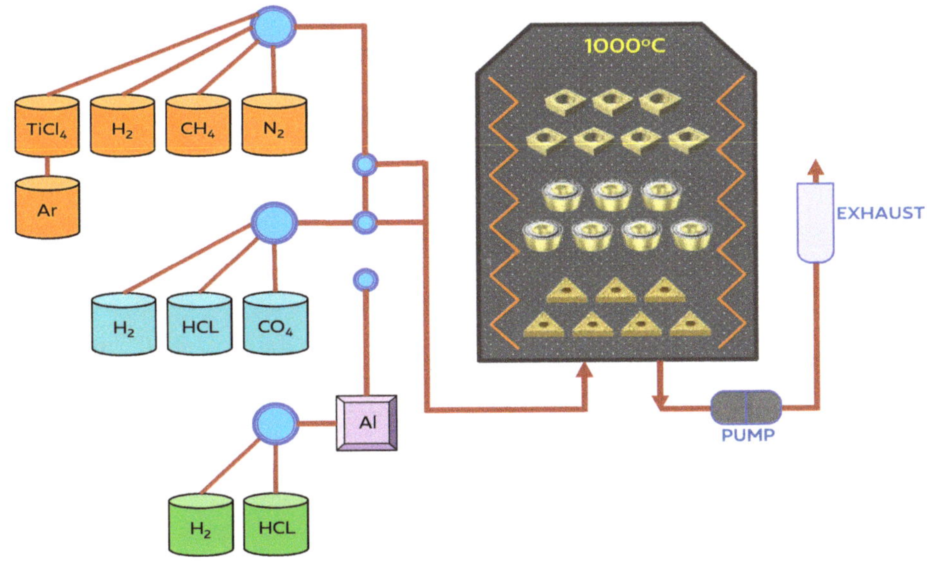

Gasses carrying Ti, Al, N, C, and O enter furnace to react on insert surfaces

REACTION FOR MODERATE TEMPERATURE CVD: ~850 - 900 ° C

$$9H_2 + 6TiCl_4 + 2CH_3CN \rightarrow 6Ti(CN) + 24HCl$$

REACTION FOR HIGH TEMPERATURE CVD: ~950 - 1050°C

$$2TiCl_4 + 4H_2 + N_2 \rightarrow 2TiN + 8HCl$$

$$TiCl_4 + CH_4 \rightarrow TiC + 4HCl$$

$$2TiCl_4 + N_2 + 2CH_4 \rightarrow 2TiCN + 8HCl$$

[**NOTE:** With all the HCL produced in the above reactions, much attention needs to be given to maintenance to prevent corrosion.]

Figure 26. Chemical Vapor Deposition (CVD) – Application Process.

CVD Advantages & Usage

- CVD is better suited for edges that are honed.
- CVD offers excellent adhesion of coating.
- CVD offers good coating thickness and uniformity.
- CVD offers a variety of coating layering options.
- CVD helps protect carbide substrate from abrasive wear and heat.
- CVD is not a line of sight process, and therefore all features coated.
- There are no compressive stresses generated in CVD.
- Best used when:

- Wear resistance is of primary importance.
- Roughing operations.

4.3.6.3 MODERATE TEMPERATURE CHEMICAL VAPOR DEPOSITION (MTCVD)

More important, and aligned with the mission of this text is the understand of the limitations and correct application of the MTCVD coatings:

MTCVD Process Highlights

- MTCVD is a higher temperature (lower than CVD) process (700 - 1100°C).
- MTCVD is a temperature driven reactive gas process.

MTCVD Coating Materials

- TiN, TiCN, & TiC.
- Al_2O_3.

MTCVD Limitations

- MTCVD creates some substrate strength loss.
- MTCVD requires edge honing.
- Not all material compositions are available at low end temperatures.

MTCVD Advantages & Usage

- MTCVD offers excellent adhesion of coating.
- MTCVD offers good coating thickness and uniformity.
- MTCVD offers a variety of coating layering options.
- MTCVD affects substrate less than the high temperature CVD.

4.3.7 PVD VS CVD COATING COVERAGE

In closing, ***Figure 27*** shows one of the primary differentiation in coverage between the PVD & CVD process, especially in terms of coating coverage. PVD being a 'line of sight' process simply coats everything in front of it. The CVD process coats by way of particulates filling up a chamber, it is able to get fuller coverage all around the object. As seen in ***Figure 27***, the PVD process occasionally leaves void regions uncoated, which could create potential breakpoint of the coating. Uncoated regions are acceptable in regions of the insert (vis. Screw Holes) that do not detrimental affect the cutter performance.

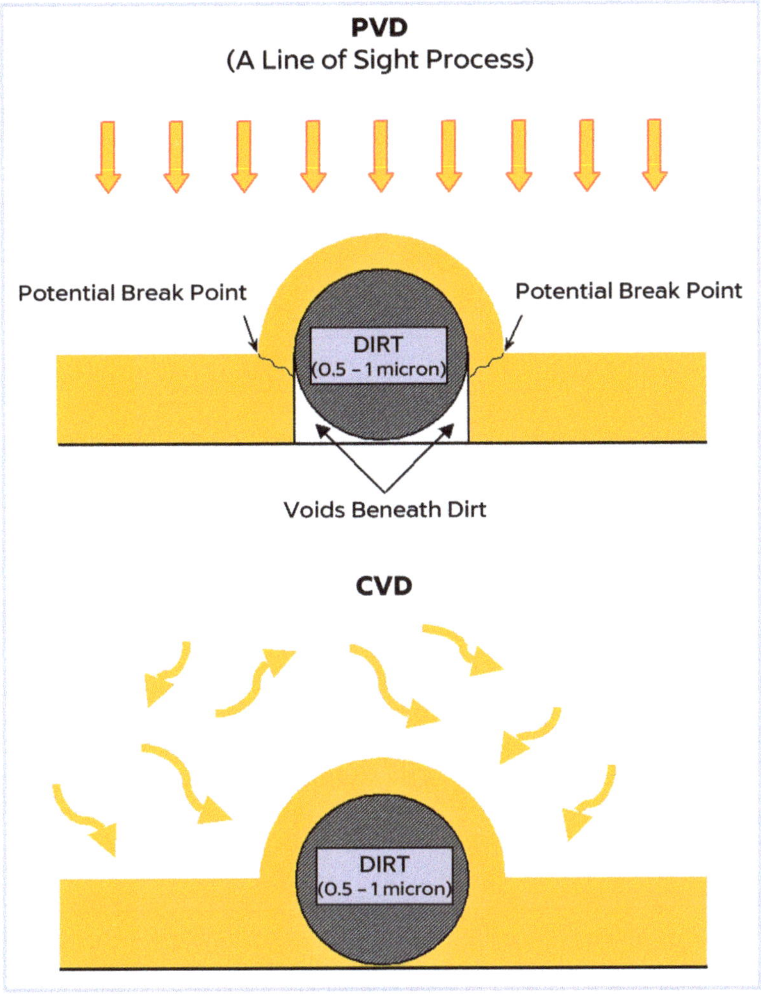

Figure 27. Mechanism of PVD & CVD – Coating Coverage.

5

CUTTING TOOL ASSEMBLY NOMENCLATURE

In modern day CNC operations, the degree of attention to detail in correctly assembling a cutter in a holder can make the difference between a seamless operation, and catastrophic failure. In this section, we will discuss some key features and nomenclatures commonly used by industry in the assembly of cutting tools & holders.

5.1 THE HOLDER

At the very fundamental level, a holder can be partitioned into two distinct sections, ***Table 1***: First is the taper section. This is the section of the holder that goes into the machine spindle. Second, the tool holding section. This is the section of holder responsible making the connection with the cutting tool. The techniques to do either task are numerous, and will be discussed in a later section. However, there are some key nomenclatures that need to be understood at this point:

The Gage Line: Most holders have a feature known as the Flange. One of the main functions of the flange is to facilitate a holding feature that allows the tool change gripper arm to hold the tool-holder assembly and move it from the tool chain in a machine to the spindle. As seen in ***Figure 28***, the back of the flange face mates up with the front of the spindle. This location forms an imaginary line called the Gage Line, and is the critical starting point for most assembly measurements.

The Holder Length: Is simply the length of the holder from the Gage Line.

The Grip Length: Particularly in regards to Solid Carbide tools, this is the maximum length that the tool can be embedded in holder.

A **Best Practice** in terms of Grip Length is that it should be at least 2 times the diameter of the Tool that is being held.

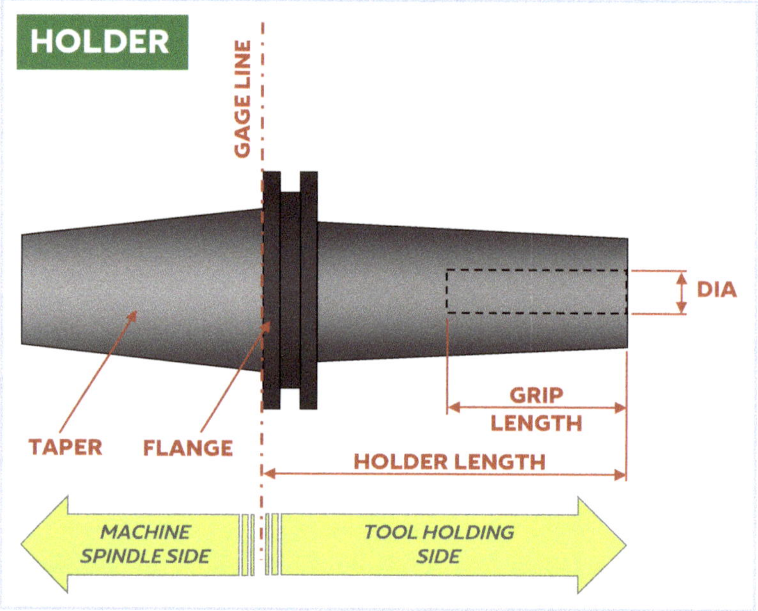

Figure 28. The Holder Nomenclature.

5.2 ENDMILL & DRILL ASSEMBLY NOMENCLATURE

Figure 29 shows the key assembly nomenclatures for a typical Endmill and drill assembly, ***Figure 30*** shows that of Facemill & Slotter, and ***Figure 31*** for a Right-Angle Head.

The Flute Length: Is the maximum effective functional cutting length of the Endmill or drill. Simply put, the tool cannot be engaged in a workpiece at lengths any longer than the Flute Lengths.

The Clear Length: Is also known as Clearance Length. As the name suggest this is the total length of the tool that is cleared to enter into the workpiece without causing any collisions. Usually a user should observe Clear Length as the length from the tip of the cutting tool to the first abrupt change in geometry along the cutting length of the cutting tool or holder that will cause a detrimental interference with the workpiece.

The Gage Length: Is arguably the most critical dimension required in a cutting tool and holder assembly. Simply put it is the dimension from the Gage Line to the tip of cutting tool. When a user assembles a cutting tool to the holder, a key dimension that they are working to is the Gage Length. As in most tool and holder assembles there is opportunities to make micro adjustments to the Gage Lengths so that the assembly comes out to exactly the specified prints. Getting this dimension wrong has been a root cause of many a machine crashes and catastrophic failures.

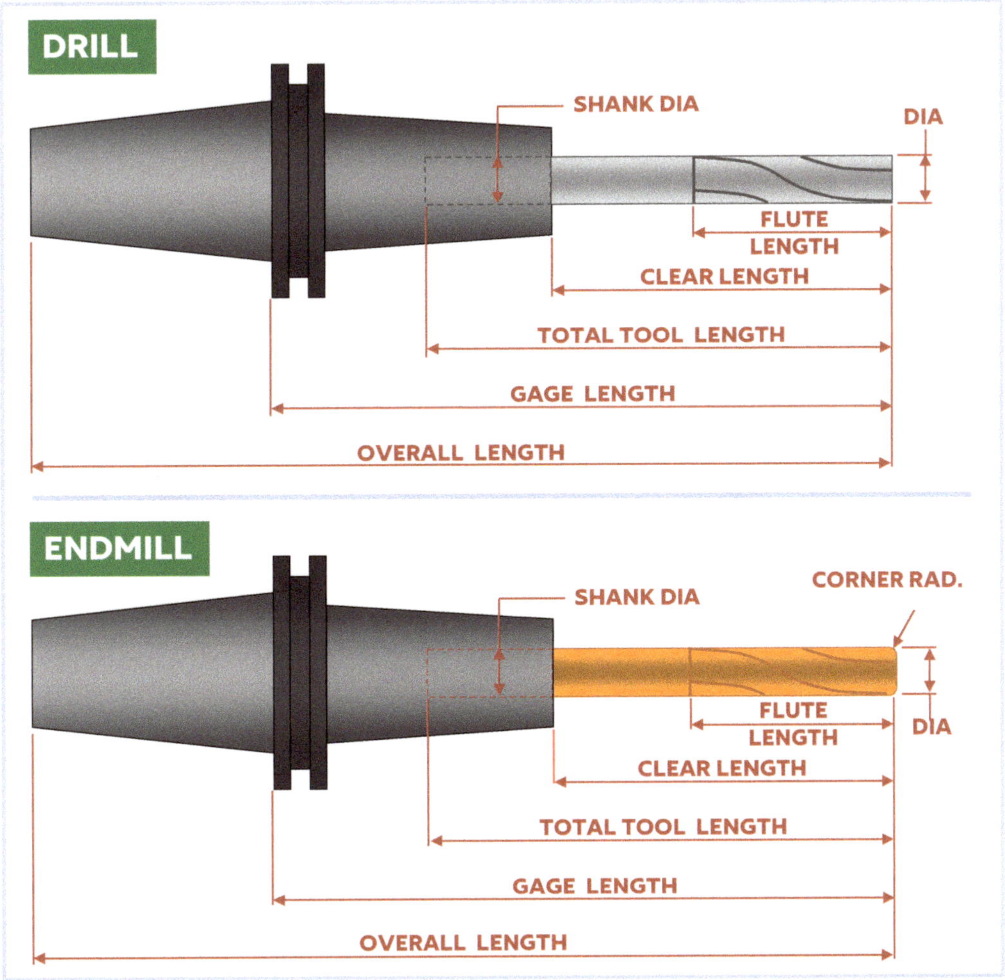

Figure 29. Cutting Tool & Holder Assembly Nomenclature (Drill & Endmill).

The Overall Length: Is the dimension from the tool tip to the very back on of the holder. The Overall Length along with Gage length become important when it comes to making sure that we are creating an assembly that fits in the max length limits of the machine's tool chain or carousel.

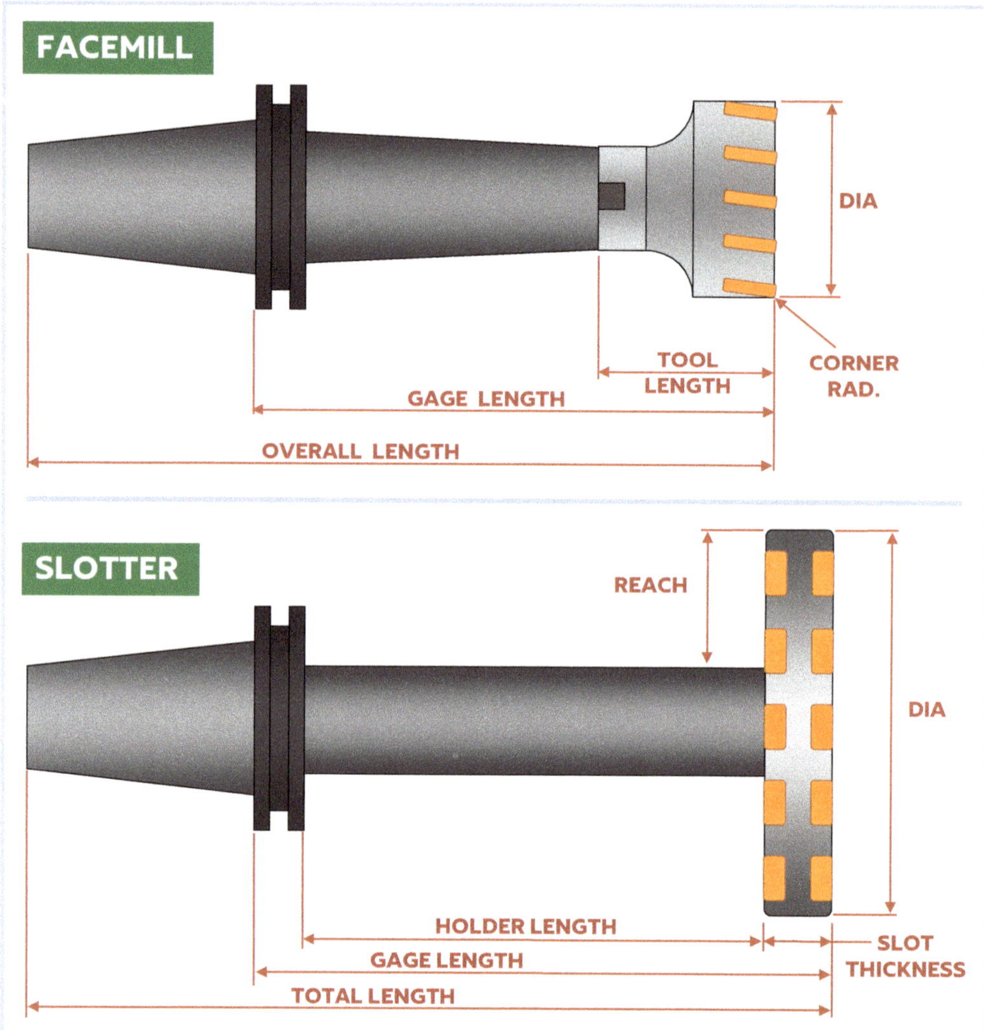

Figure 30. Cutting Tool & Holder Assembly Nomenclature (Facemill & Slotter).

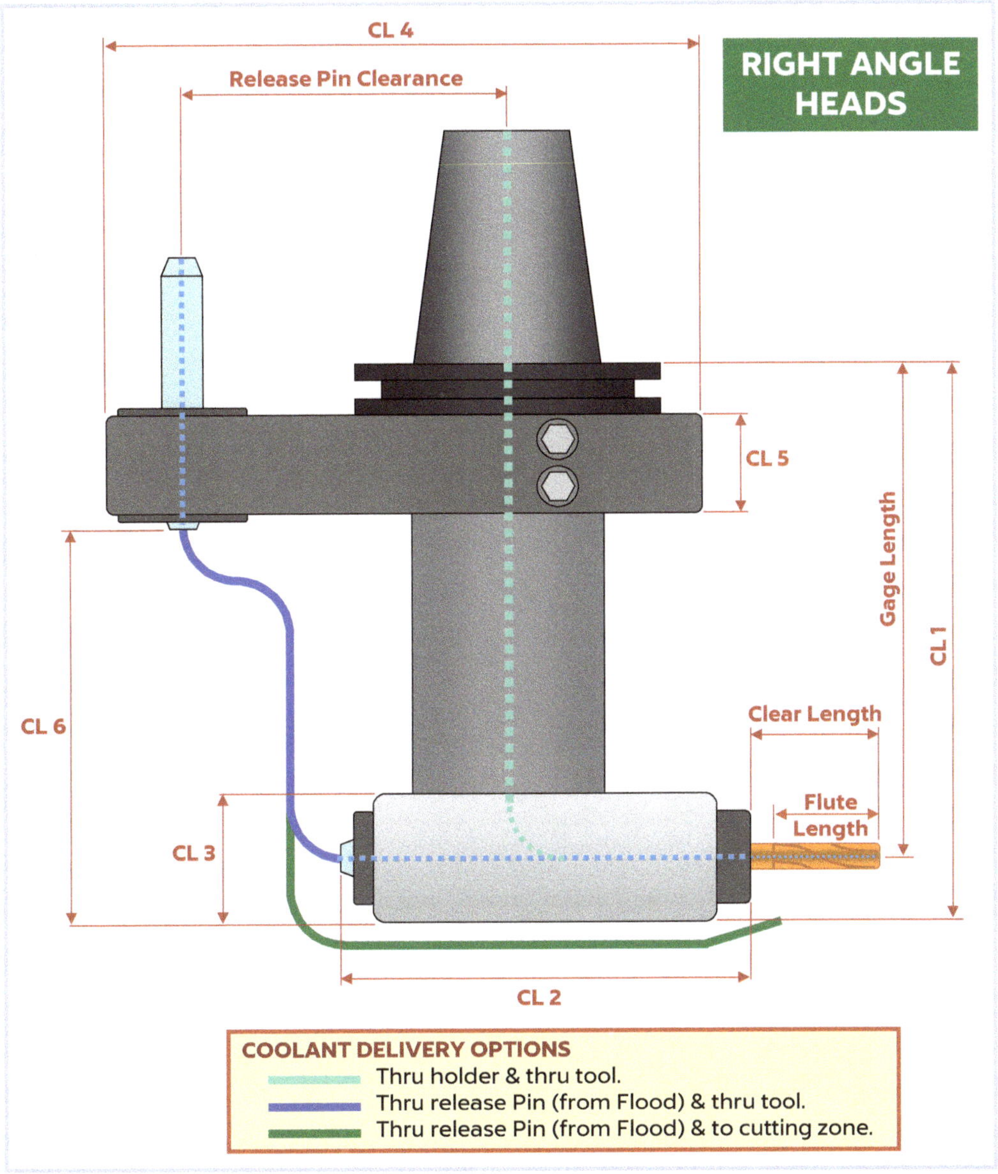

Figure 31. Cutting Tool Assembly Nomenclature (Right Angle Head).

6

TOOL HOLDING SYSTEMS

Now that we have discussed tool & holder assembly nomenclature in the earlier chapter, it is a good idea to understand the diversity of types and sizes of Tool Holders. It is also important for the user to have a high-level understanding of Holders before we talk about topics such as Bending Moment Limits, Balancing, and Vibrations in metal cutting. The intent of this chapter is not to cover the full spectrum of every single holder that it out in the market, but rather take the approach of focusing in on key holder related parameters to help the user make informed choices.

A Tool Holder is basically an interface between the cutting tool and the machine's spindle. We certainly do not want this connector to become the weakest link in the cutting operation. Investments in high performance spindles, and state of the art cutting tools can be rendered useless if proper attention is not given to selecting the correct Tool Holding System. When holding systems are overlooked, the end-user often ends up accommodating the limitations of the holder by way of: dialing down machining processes, slowing down Speeds & Feed, underutilizing spindle Power & Torque. Once again, for the learnings in this chapter on Tool Holding Systems we will take the approach of "Preventing bad things from happening".

6.1 A TOOL HOLDER BASICS

As seen in ***Figure 32***, the best way to visualize a Tool Holder is to partition its features into two important sections:

- The Machine Spindle side (aka. The Taper Side).
- The Cutting Tool side.

The line that partitions these two sections is known as the Gage-Line, ***Figure 32***. The Gage-Line is located where the Flange of the holder mate's up to the machine's spindle.

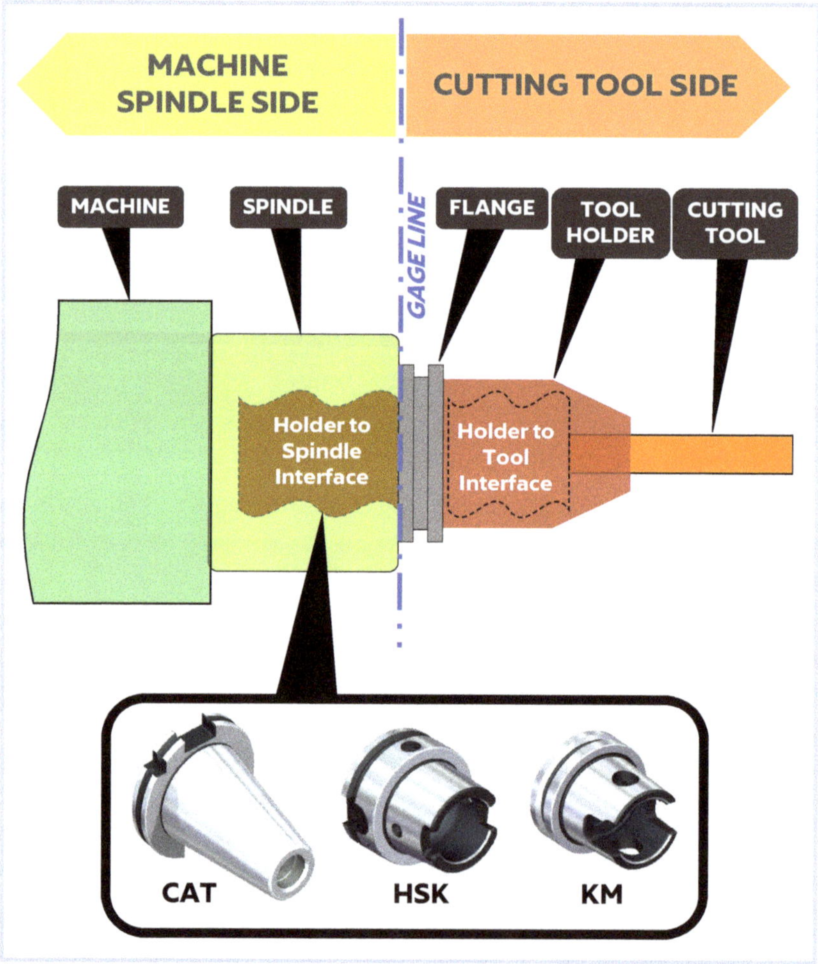

Figure 32. Tool Holder Basics.

6.2 TOOL HOLDER – THE TAPER SIDE

As seen in ***Figure 32***, the Taper side of the holder is that part that goes inside the machine's spindle. At a high level, the machining industry has two dominant Taper Types (***Figure 33***), each with their various sizes:

- The CAT taper System.
- The HSK taper System.

It should be noted that to further a further feel of the relative masses of equivalent performance systems, the schematics in ***Figure 33*** are drawn to their relative scales (Here CAT50 vs HSK100).

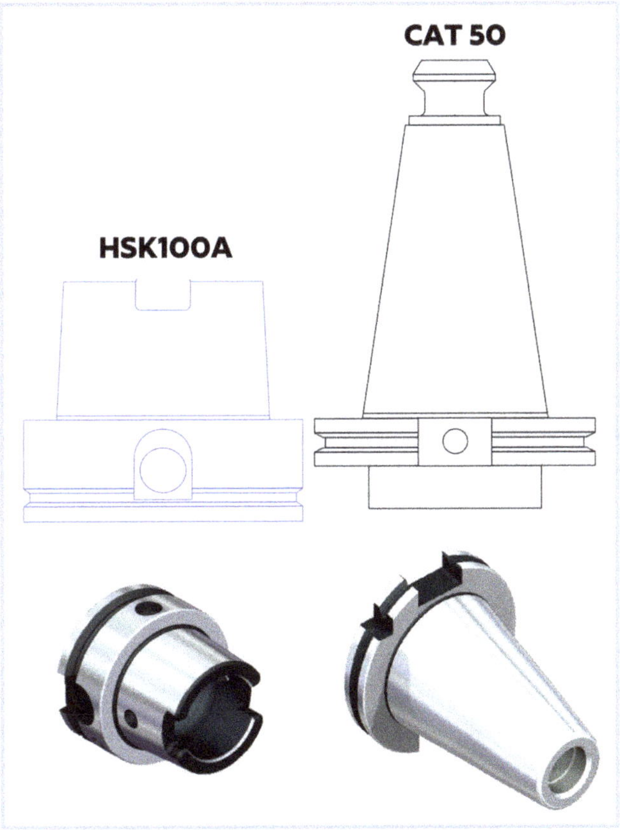

Figure 33. The CAT vs HSK Taper Systems.

NOTE: In this text we will commonly compare the HSK and CAT system by referring to their **Equivalent** sizes. It is important to clarify the term '***Equivalent***'. This basically refers to the standard offerings of Spindle Taper sizes by machine builder. These spindle tapers are considered close enough in size, and cost of offerings. Additionally, they are also driven out by way of equivalent machine footprint, and spindle size. For example: When machine builders propose a spindle size to a customer, they may suggest a CAT50 side by side to an HSK100 for consideration. This does not necessarily mean that the holders are equivalent in performance.

6.2.1 THE CAT TAPER SYSTEM

The CAT Taper system has its roots from the Caterpillar Company, and hence the name. As seen in ***Figure 33***, this is a conical type tape that is pulled into the spindle when the machine's draw-bar grips and pulls on the holder's retention knob, ***Figure 34***. Special attention is given by holder manufacturers to precisely tolerance not only the taper size, but also its relative position with the flange. As can be seen the CAT system has substantially more amount of mass of steel as compare to the equivalent HSK system.

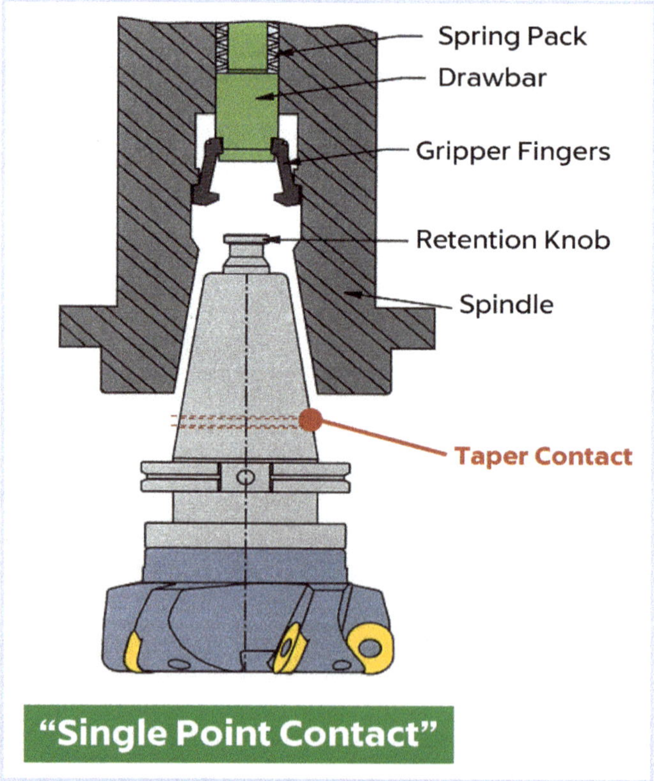

Figure 34. The CAT System.

6.2.1.1 SIZES OF CAT TAPER SYSTEM

The CAT taper is available in number sizes: 30, 40, 45, 50, and 60. Each size has a published performance band which will be discussed in subsequent sections.

6.2.2 THE HSK TAPER SYSTEM

The abbreviation **HSK** for "Hohl Schaft Kegel," which, literally translated into English, means "hollow-shank taper." As seen in ***Figure 33***, the HSK is a hollow and gets pulled into the machine spindle with gripper bar, ***Figure 35***. The grippers go inside the hollow taper and execute a pull as they spread out. The HSK is lighter and more symmetrical system than its equivalent CAT.

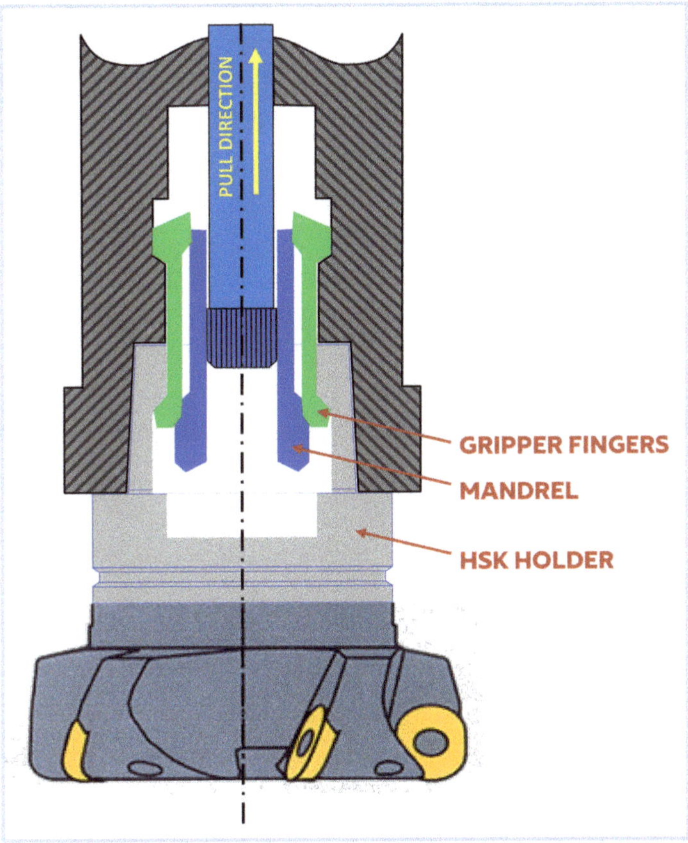

Figure 35. The HSK System.

6.2.2.1 SIZES OF HSK TAPER SYSTEM

The HSK taper is available in number sizes: 63, 80, 100, 125, and 160. Each size has a published performance band which will be discussed in subsequent sections.

6.3 TAPER TYPE & SIZE SELECTION CRITERIA

Every tool holding system and its size comes with its set of capabilities and limitations. The key criteria and performance specs for holders are:

- Maximum operating RPM.
- Maximum Bending Moment Limit.
- Maximum Clamping Force from Drawbar or Gripper.
- Retention Knob Torque Setting.
- Machine Footprint.

The values for each of these criteria are summarized in the table shown ***Table 2***.

TAPERS	MAX. RPM	MAX. BENDING MOMENT		MAX. CLAMPING FORCE		RETENTION KNOB TORQUE
	(rev/min)	(Nm)	(lbf-in)	(kN)	(lbf)	(ft-lbf)
HSK63	28,000	420	3,717	18	4,047	NA
HSK80	22,000	810	7,169	28	6,295	NA
HSK100	15,300	1,250	11,063	45	10,116	NA
HSK125	10,000	2,900	25,667	75	15,737	NA
HSK160	7,000	3,616	32,000	115	25,853	NA
CAT30	14,000	203	1,800	5	1,200	8
CAT30 SIMUL	14,000	294	2,600	5	1,200	8
CAT40	10,000	390	3,450	10	2,300	30
CAT40 SIMUL	10,000	542	4,800	10	2,300	30
CAT45	8,000	678	6,000	18	4,000	50
CAT45 SIMUL	8,000	915	8,100	18	4,000	50
CAT50	6,000	847	7,500	22	5,000	75
CAT50 SIMUL	6,000	1,096	9,700	22	5,000	75
CAT60 (All Flanges)	3,600	2,147	19,000	58	13,000	120
CAT60 SIMUL (139 mm Flange)	3,600	2,576	22,800	58	13,000	120
CAT60 SIMUL (150 mm Flange)	3,600	2,757	24,400	58	13,000	120
CAT60 SIMUL (200 mm Flange)	3,600	3,582	31,700	58	13,000	120

Table 2. Tool Holding System – Cross Reference Table.

6.3.1 MAXIMUM RPM LIMITS

The maximum RPM is the highest speed each of these holders can be allowed to operate at, ***Table 2***. From this viewpoint the user should consider the following factors in their production application:

- Workpiece material being machine: Is the metal hard or soft? Consequently, the RPMs will need to calculated for each cutter DIA based on fixing the SFM for that material.
- Machining Strategy: Is the process going to have an Aggressive Roughing strategy. (ie lower SFM), or is it going to have a High Velocity strategy (Higher SFM)? Each of these will dictate the upper end of RPMs required, and consequently the holder selection.
- Type of Cutting Tool: What is the material and type of cutting tool utilized? Are the cutting tools state-of-the-art modern-day coated carbide, or are they earlier generation HSS / Cobalt cutters? This could even be decision based on a shops budget for tools. This decision would drive out the SFM, and consequently the RPM required.

NOTE: Typically, the HSK systems are capable of accommodating higher RPMs as compare to their equivalent CAT systems.

6.3.2 MAXIMUM BENDING MOMENT LIMITS

In a future chapter we learn about calculating Bending Moment B_M in a cut. However, for the sake of the text it is important to introduce this concept while discussion holders.

$$B_M = F_T \times \left[G_L - \left(\frac{DOC}{2} \right) \right] \qquad [Units: in.\, lbs]$$

Where F_T is the total cutting force, G_L is the Gage Length, and DOC is the axial depth of cut. The primary purpose of this calculation is to help determine the ideal Tool Holding system for the application, ***Table 2***. This is a very critical calculation, and one that is often overlooked. If the Total Cutting Forces are too high, this can lead to a Bending Moment that exceeds the limit of the holder type or size. This consequently causes micro-separation between the holder and spindle face. It eventually manifests itself as a vibration. In the worst-case scenarios, exceeding the bending moment limit has led to cracking of the holder walls, and even catastrophic failure where the holder falls out from the spindle.

NOTE: Typically, the HSK systems are capable of accommodating higher Bending Moment as compare to their equivalent CAT systems.

6.3.3 MAXIMUM CLAMPING FORCE FROM DRAWBAR OR GRIPPER

The maximum clamping force is the amount of force with which the holder is pulled into the spindle, ***Table 2***. This value is provided here for user reference only. This is basically a typical performance spec of the machine's drawbar. Machine builders may vary slightly on the amount of force with which they choose to pull on the holder.

There are other benefits to knowing the maximum pull force on the holder. As the machine cycles through years of production, the amount of force with which the holder is pulled into the spindle tends to drop off. This causes micro-separation between the holder and spindle face, and can manifest itself by way of vibrations during cutting. As a preventative maintenance, it is a good idea to check the drawbar pressure regularly.

On the other hand, pulling the holder into the machine spindle with excessive force can be detrimental as well. Excessive pull force can deform or crack the HSK holder wall, and stretch or crack the retention knob on a CAT system. Additionally, excessive pull force can lead to a holder not releasing from the spindle face when it is done cutting. Such situations have been known to turn into a machine maintenance nightmare.

6.3.4 THE RETENTION KNOB TORQUE SETTING

If the user chooses to use a CAT system it will use a retention knob as compared to an HSK. The value of the retention torque setting is for user reference only. Considering the importance of the work this knob is doing, it is a best practice to torque it to its published spec and avoid any catastrophic failures.

6.3.5 THE MACHINE FOOTPRINT

Sometimes the decision for taper type and size boils down simple to the amount floor space available in the production environment. As the taper sizes get larger so do the spindles. Larger spindles require larger columns, housing, and axis drives. This consequently drive up the size of the machine's footprint. Often times users will pick a smaller holder size just so as to fit the corresponding machine offered in the available space. In such case we simply make the most of the situation, and try to apply to best machining strategy so as to optimize the process.

6.4 TOOL HOLDER - THE CUTTING TOOL SIDE

As seen earlier in ***Figure 32***, the cutting tool side of the holder is the part that grips the tool. In modern day industry, the options to grip the tool are quite diverse. For the sake this text, only a high-level attention is given to recognizing the major tool gripping systems. Special attention is given only when certain tool holding system need to be recommended for specific applications.

6.4.1 SHRINK-FIT ADAPTER

As seen in ***Figure 36***, a shrink-fit adapter works on the principle of shrinking the holder onto the tool shank. The outer wall of the holder is heated up on a specially built 'Heat-Shrink' machine. The cutting tool with a smooth shank is then dropped into the ID of the holder. While the holder is still hot, the operator may have the option to adjust the height of the tool. Eventually the holder shrinks onto the cutting tool. This cooling process is often sped up by way of using cooling sleeves.

A shrink-fit tool and holder assembly is one of the most symmetrical setups. The shrink-fit holders can provide a runout of less than around 0.0007". This makes it very ideally suited for applications that need high RPMs and balance. They however do possess the inherent threat of tool pullout. As will be seen in a later section on *tool pullout prevention systems*, design improvements have significantly mitigated this risk.

Best Practices

- When it comes to shrink fit holders, it is best to have the grip length (ie the amount of tool shank that is in the holder) be at least 2 X DIA.
- Try to use straight shank tools for shrink fit, as compared to those with Weldon flats or other features which can compromise the grip surface area.
- Shrink-fit holder are available in regular wall and heavy wall thickness. If the operation & clearance permit, consider using heavy walled shrink-fit holders as much as possible. This can become especially important in hard metals machining.
- Shrink-fit works on pretty tight tolerances between the tool shank and the holder ID. Make sure that there is no coating on the part of the tool shank that needs to go into the holder.

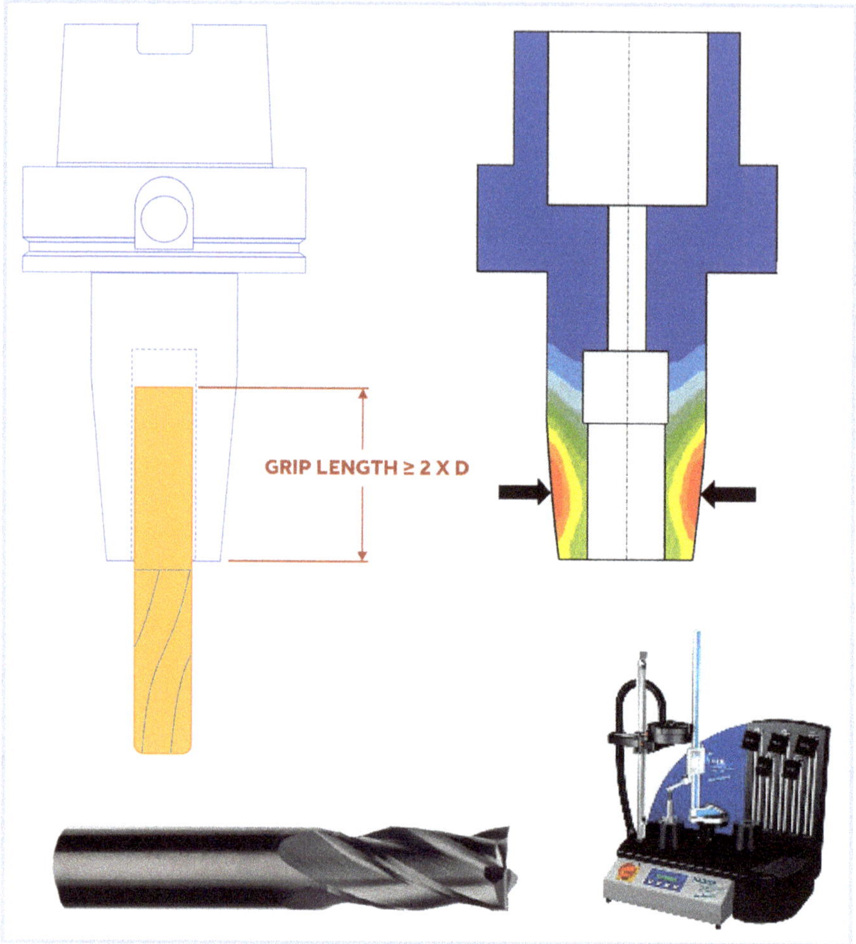

Figure 36. Shrink Fit Adapter.

6.4.2 WELDON FLAT ADAPTER

The Weldon Flat adapter as shown in ***Figure 37***, is a method of assembling the tool in the holder by way of clamping with setscrews. The Weldon Flat adapter is a simple tool holding system that does not requires any ancillary equipment like that in a heat-shrink system. The Weldon Flat system is ideal for cuts that are very aggressive and draw high forces. It is however not the most symmetrical system (due to unbalance from the set screws), and therefore not ideal for high velocity machining. Typically, the Weldon Flat holders tends to have a runout of less than 0.007". It should also be noted that there is no opportunity to adjust the height of the tool in the Weldon Flat adapter system.

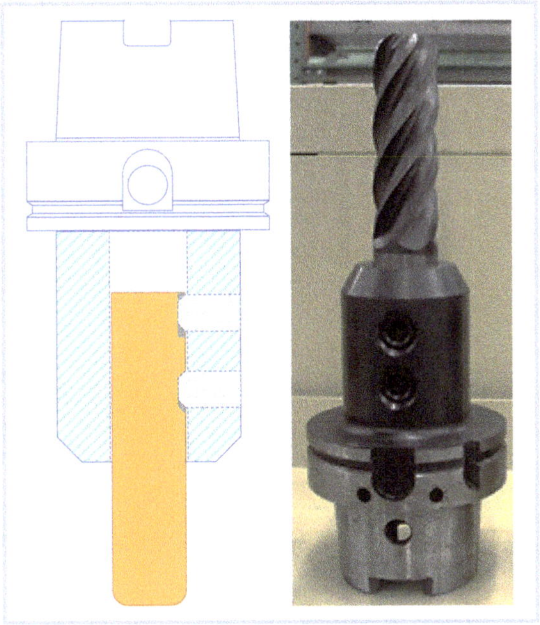

Figure 37. Weldon Flat Adapters.

The big advantage of Weldon Flat holders is that is assembled correctly there can be no tool pullout. There is however a correct way to setup a Weldon Flat assembly as shown in ***Figure 38***. Once again, the approach taken is "To stop bad things from happening". Recognize that it is better if during cutting something goes wrong, it is safer if the tool pushes into the holder, rather than pulls out of it. If the tool pushes into the holder, it will simply not cut the part as intended. However, if the tool pulls out of the holder, it can severely damage the part and even cause catastrophic failure.

STEP 1: As shown by the steps in ***Figure 38***, once the tool is inserted into the holder, start by starting the set screw #1 until it makes contact with the flat on the tool. Do not tighten this set screw at this point.

STEP 2: Next, pull the tool down (ie away from holder) so that the set screw makes contact with the rear chamfer of the flat.

STEP 3: Start the set screw #2 until it makes contact with the flat on the tool, and even tighten it to its published torque setting. Finally, tighten set screw #1 to its published torque setting.

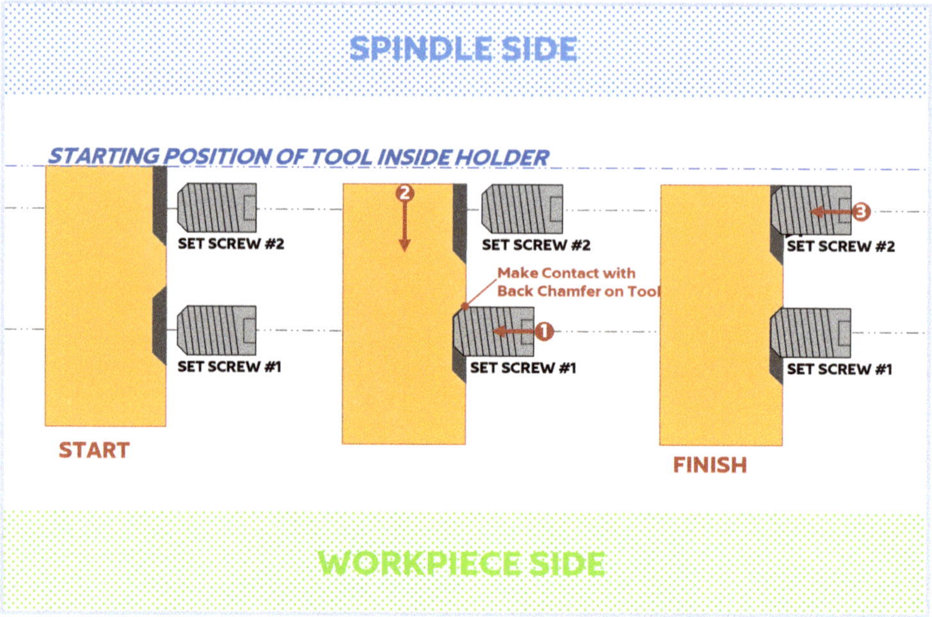

Figure 38. Weldon Flat Adapters – Correct Setup.

6.4.3 WHISTLE NOTCH ADAPTER

The Whistle Notch adapter as shown in ***Figure 39***, is a very simple system assembly by way of a set screw onto a whistle shaped (angled) flat on the tool shank. This system is suited for use in aggressive roughing. However, it is not a very symmetrical system, therefore not very suited for high velocity applications.

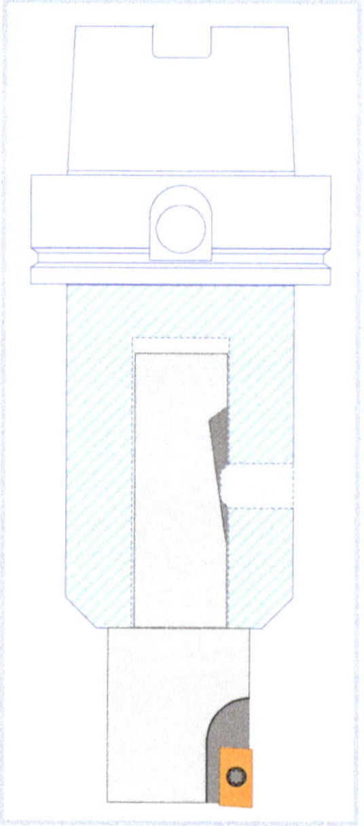

Figure 39. Whistle Notch Adapter.

One of the biggest advantages of the Whistle Notch adapter system is that the height of the tool sticking out from the holder can be easily adjusted. This adjustability become quite essential in multi-spindle machines such as gantries. Gantries cut multiple similar parts at the same time. Therefore, the tool tips have to be set precisely in height. Earlier generation gantries do have much flexibility to adjust individual spindle heights. Here the Whistle Notch system is optimal for easily fine tuning the heights of the tools across the gantries.

6.4.4 DIRECT MOUNT SHELLMILL ADAPTER

A direct mount Shellmill adapter is one of the most basic mounting systems available, ***Figure 40***. Simply put, the cutting tool is bolted on. Here the cutting tool is properly centered onto the holder by way of a 'Pilot', and then fastened by a pattern of bolts. Torque is transferred to the cutting tool through the keyways. This adaption is very common for indexable tools, especially Facemills. It is ideally suited for applications that require a high transference of Torque.

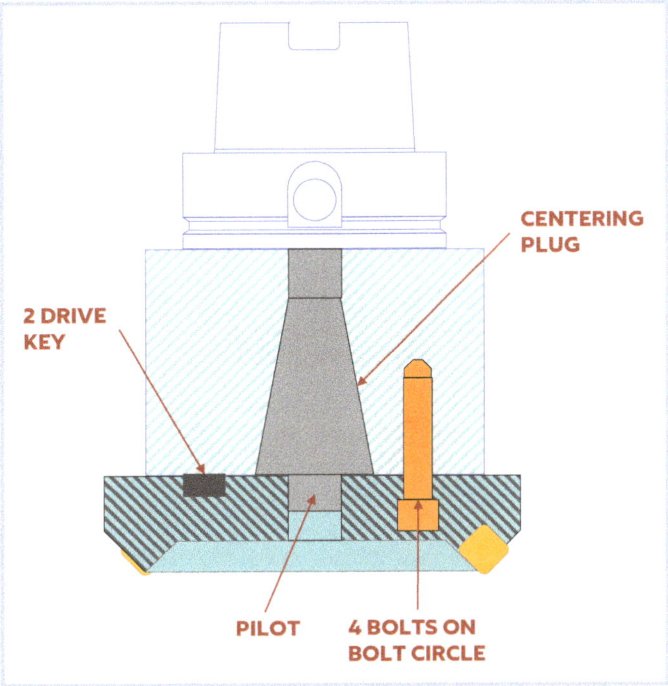

Figure 40. Direct Mount Shellmill Adapter.

6.4.5 SLOTTING WHEEL CUTTER ADAPTER

A typical slotting wheel cutter adapter is shown in ***Figure 41***. It is a very unique adaption in the sense that they tend to be strictly deducted to use in slotting wheels. Since slotting wheels can get extremely large, special attention is given to the tolerances of its adapter. This system is especially applicable for transference of extremely high torque values. The cutting wheel is centered on to the arbor and mounted either by way of a single bolt, or by way of a bolt pattern (on larger wheels). The torque is transferred to the cutting tool by way of drive keys.

For larger slotting wheels, these adapters are typically custom made. Attention is also given to the size and number of bearings inside some of these arbors. Additionally, planning the coolant deliver through the arbor without losing much pressure can also be quite an engineering feat.

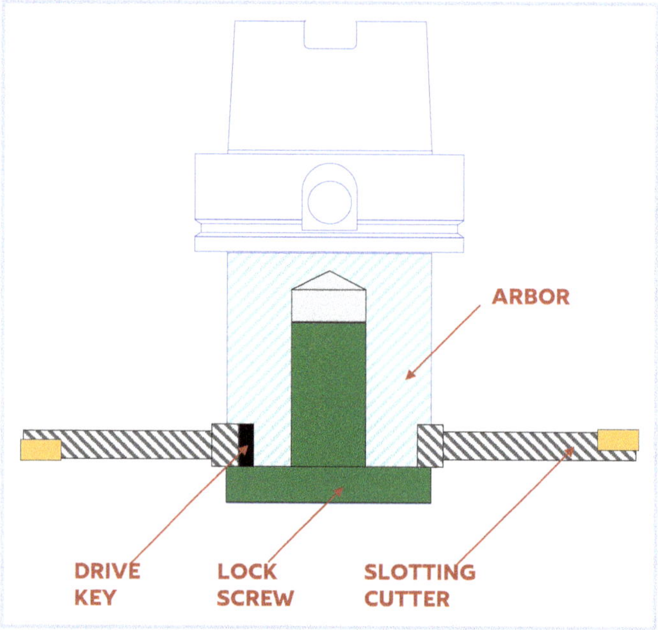

Figure 41. Slotting Wheel Cutter Adapter.

6.4.6 HYDRAULIC HOLDER ADAPTER

As seen in ***Figure 42***, a hydraulic holder adapter is very much similar to a shrink-fit type clamping system, only with the need for heating or cooling the adapter. The hydraulic adapter works on principles of fluid mechanics where in the user simply turns on the adjusting setscrew to initiate the input pressure. This in turn magnifies the hydraulic pressure inside the holder near the tool gripping area. Specially designed metallic sleeves or bladders then press onto the shank of the cutting tool.

There are some distinct advantages to the hydraulic adapter. In recent years, significant advancements have been made in hydraulic holders to where they can meet, if not even exceed the grip strength of a shrink fit holder. Additionally, various smaller sleeves can be dropped into the ID of the holder to accommodate multiple sized tools. Thus, making this system quite versatile. Hydraulic adapters tend to provide inherent dampening from the holder mass and its hydraulic fluid.

They are very symmetrical systems, and well suited for high velocity applications. Perhaps the biggest advantage of the hydraulic adapter is the safety factor created by the elimination of heating and cooling of the holder.

Hydraulic adapters do tend to be bulkier in diameter then their other counterpart holders. User should be mindful of clearing related criteria while selecting this system.

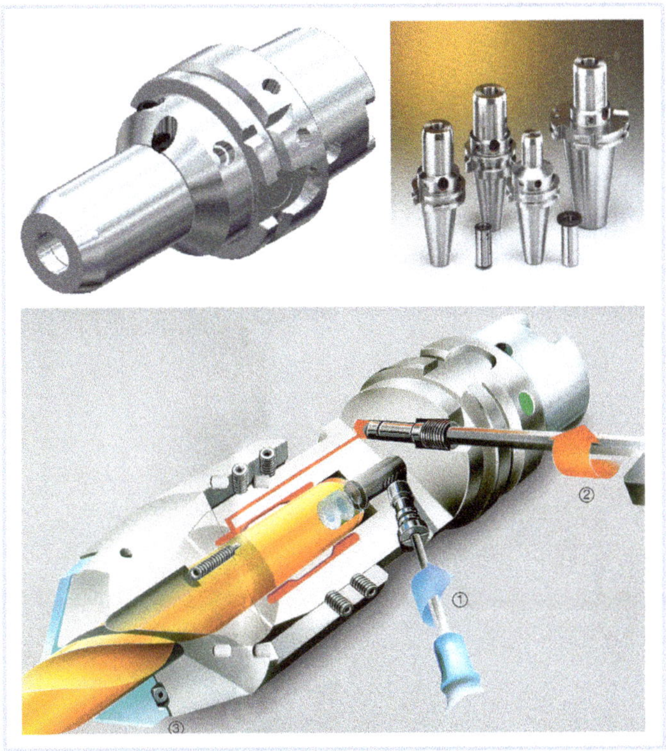

Figure 42. Hydraulic Holders.

6.4.7 COLLET CHUCK ADAPTERS

A collet chuck adapter system is shown in ***Figure 43***. A collet can be look at as a precision sleeve. This sleeve is designed with numerous relief type cuts on its OD which allow it to be pinched in. Tightening the locknut onto the holder pinches in the collet, which then clamps on to the cutting tool. Collet chuck adapters are well suited for lighter cuts with smaller endmills. Excessive forces tend to pull out cutter from collet chucks. That said, collect chuck are well suited and popular for use in drilling and reaming operations of all sizes.

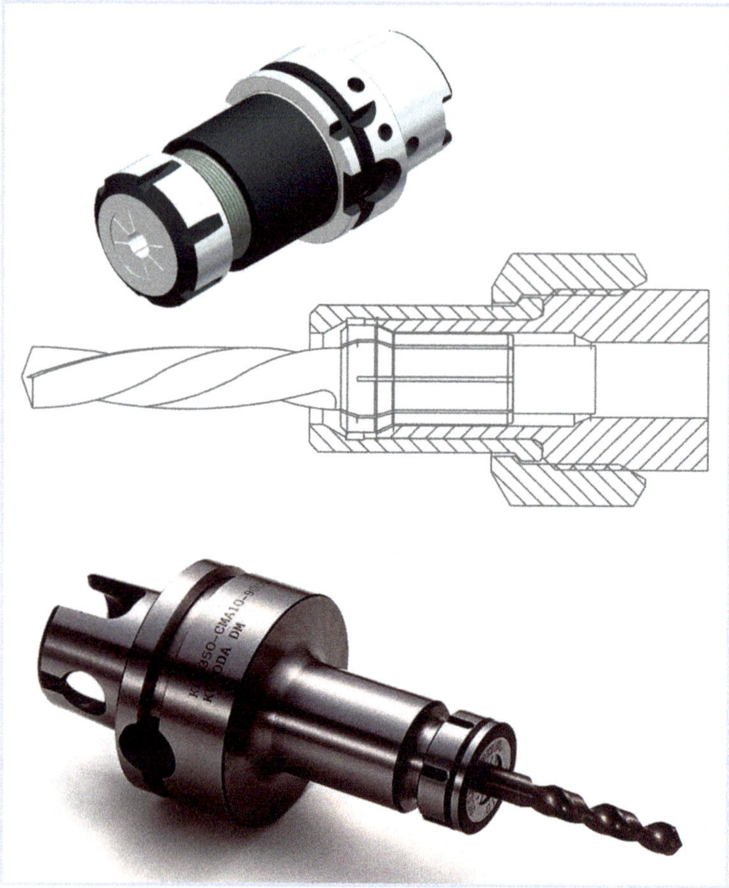

Figure 43. Collet Chuck Adapters.

6.4.8 TOOL PULLOUT PREVENTION SYSTEMS

There are numerous design improvements made to the shank of the tool, and the ID of holder to prevent issues around tool pullout. This pullout issues are typically experienced in shrink-fit, and hydraulic holders.

6.4.8.1 THE SAFE-LOCK SYSTEM

A popular design improvement to mitigate tool pullout issues has been the Safe-Lock system by Haimer, ***Figure 44***. Specially designed spiral grooves are ground into the shank of the cutting tool. The tool is than fitted into the holder (shrink or hydraulic), and the design feature in the holder's ID mates with the grooves. As the tool is pushed further down into the holder, the mating of grooves forces it rotate opposite to the cutting direction. What this does during actually cutting is that, the cutting forces would tend to actually push the cutter further into the holder. In other words, if the safety of the shrink-fit were to give way to the cutting forces, the tool would never pullout, but rather the cutting forces would push it into the holder.

Figure 44. Tool Pullout Prevention System – Safe-Lock.

7

THE HAND OF CUTTER ROTATION & NUMBER OF EFFECTIVE

7.1 THE HAND OF ROTATION

The hand of rotation is basically a geometrical factor that the designer of a cutting tool needs to be aware of, and engineer a cutting tool that is best suited for the direction of the machine's spindle rotation (i.e. clockwise or counterclockwise). Although most of the industry standard is to create cutters with Right-Hand Rotation, for the sake of a well-rounded learning it is important to be able to distinguish between the two. We will also identify some rare and unique applications that require a Left-Hand Rotation cutting tool.

7.1.1 RIGHT-HAND ROTATION

As seen in ***Figure 45***, if you point your right hand thumb out and away from the machine spindle and towards the tool tip, then the direction of curl for the rest of the fingers should align with the cutting direction of the edges. This cutter design is commonly referred to as a Right-Hand cutter ***Figure 46***.

7.1.2 LEFT-HAND ROTATION

As seen in ***Figure 45***, if you point your left hand thumb out and away from the machine spindle and towards the tool tip, then the direction of curl for the rest of the fingers should align with the cutting direction of the edges. This cutter design is commonly referred to as a Left-Hand cutter.

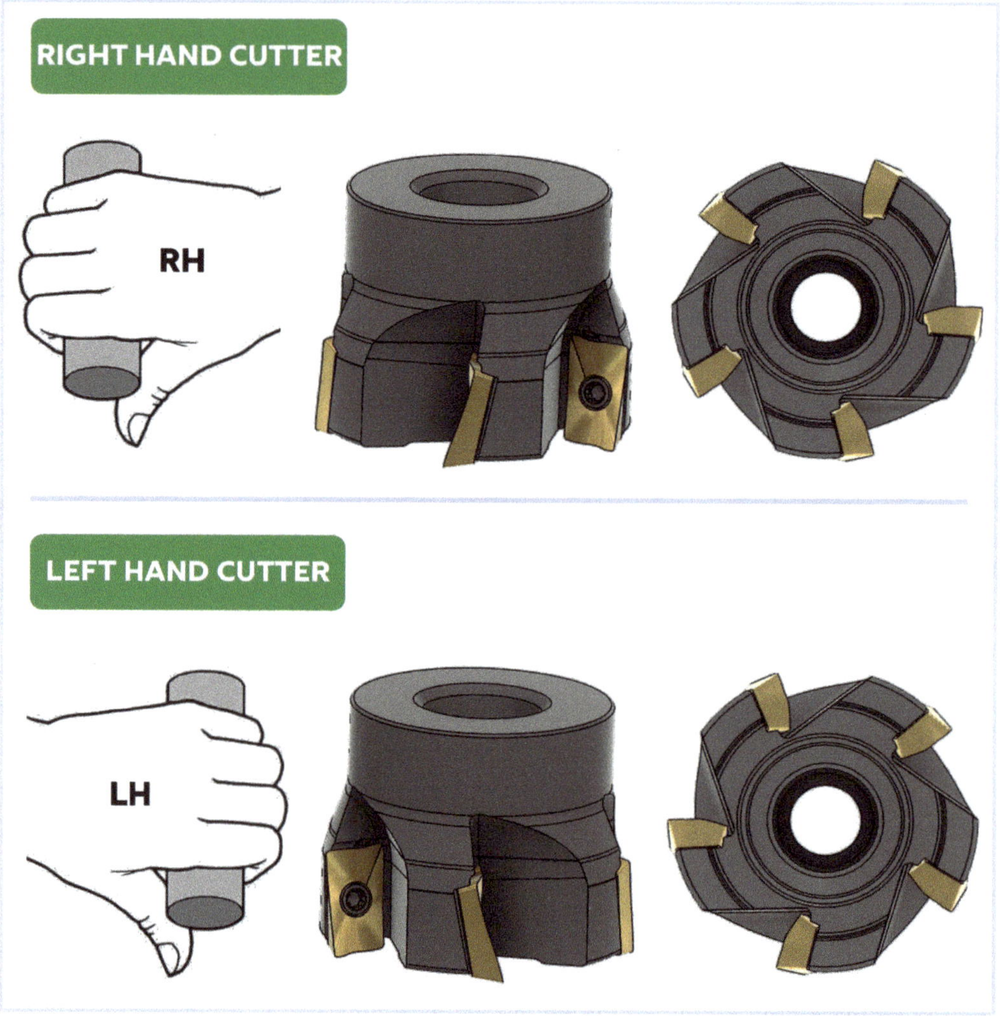

Figure 45. The Hand of Cutter Rotation.

Figure 46. Right Hand Rotation.

Although most of the cutters in the global market are universally Right-Handed, there are some unique applications where Left-Hand rotation cutters are necessary:

Screw Removals: As you can imagine most fasteners screws have a Right-Hand thread. Occasionally, due to improper usage the heads of the screws get damaged or broken and the screw needs to be removed. A commonly used trick is to use a Left-Hand drill to drill down into the center axis of the screw. The drill will usually bind up with the screw material, and will thread it out of its hole.

Right Angle Heads: As seen in ***Figure 47***, the Right-Angle Head is a customized tool holding solution used to position to tool at 90° to the spindle axis. It primarily works on the amount and type of gearing. This needs to raise a caution. Often times the Right-Hand rotation of the spindle, can yield Left-Hand rotation of the tool. This output is a consequence of the gear pairings in a Right-Angle Head. In such a case a Left-Hand Rotation tool need to be used in the Right-Angle Head.

(Of course, in the above example the user has the option spinning the machine spindle in the opposite Left-Hand Rotation, so that the output can yield the use of a standard Right-Hand Rotation cutter. Either way, the reason for this discussion is to raise awareness so that bad things don't happen in a machining operation, and that good choices are made).

Fixturing Limitations: *Figure 48* shows an example of a basic slotting application with a Right-Hand Rotation slotting wheel, that travels from left to right on the part. For the sake of this discussion, let's say that based on some geometry of engagement with the part this gives rise to some chatter or vibration type issues. It is deemed necessary that in order to eliminate the vibration the part needs to be cut from the right to left side. Now at this point you have the option to design and build a flipped fixture, but this may be significantly more expensive than the cost of a new Left-Handed slotting wheel. Although rare, this is another case where engineering a Left-Handed cutter becomes necessary.

7.1.3 HOW I WAS TAUGHT HAND OF ROTATION!!!

Although I do not recommend users actually try this technique, it may be a good **visualizing aid** to figure out hand of rotation of a cutter. Alternately, this technique may create some nostalgia amongst veteran machinists, who used this teaching aid as a source of practical jokes, indoctrinations, and ultimately much amusement. Here goes:

"Setup a cutter and holder assembly on a workbench or tool pod so that the tip of the tool is sticking up vertically. Now **visualize** safely placing each of your palms on either side of the cutting tool. Next **pretend** that if you were to take a step backward (or pull your palms towards your body). One of your palms tends to get snagged on the cutting edges and stands to get cut. That bleeding hand incidentally is also the Hand of Rotation of the cutter."

LEFT HAND CUTTER

- Used in special applications.
- Used in machinery or holding mechanisms that use gearing the reverses the direction of rotation.
- Used where limitations of spindle orientation and part fixturing require a cutter to feed in a direction that requires LH rotation.

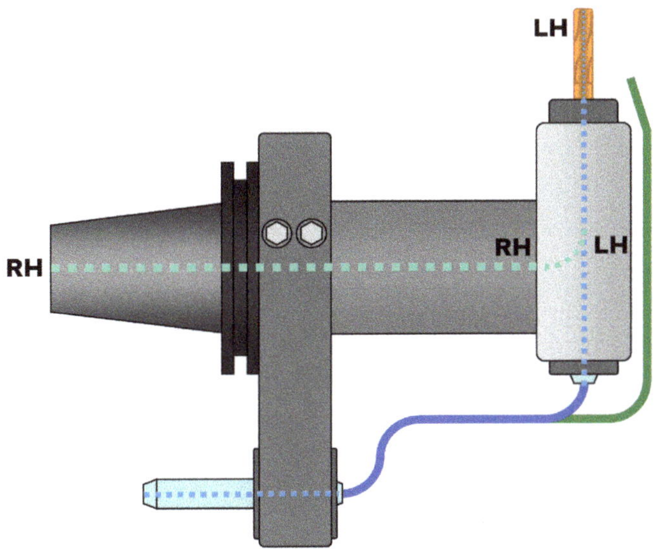

Figure 47. An Example of Use of a Left-Hand Cutter with Right-Angle Head.

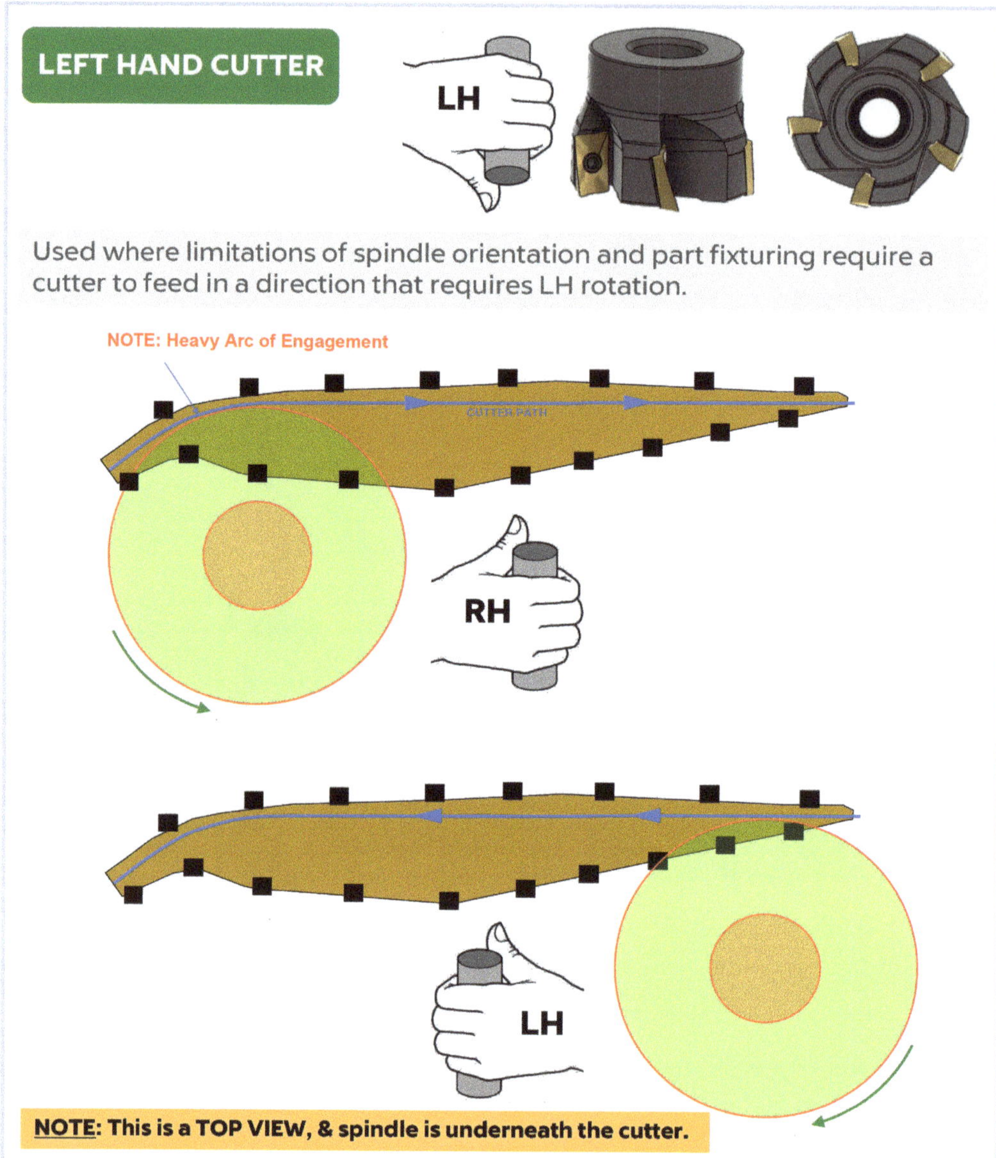

Figure 48. An Example of Use of a Left-Hand Cutter to Accommodate Fixture Limitations.

7.2 THE NUMBER OF EFFECTIVE EDGES

Visualize a child on a beach excavating a trench in the sand with its bare hands. The child fashions five of its fingers on one hand to create the shape of a shovel. Each of the fingers can be viewed as a cutting insert. However, when 5 of these fingers are put together, they collectively form One Effective cutting edge that together scrapes out a trench that spans the width of a palm. If the child were to use both of its hands (where one hand directly follows the path of the prior) to improve excavating efficiency, we would say that it now has Two Effective cutting edges (while using 10 cutting inserts, aka fingers).

Now, if the child decides to repositions its hands to where they were side-by-side and dig a wider trench, it would now have 10 total fingers collectively digging a trench that is twice as wide. In such a case 10 of these fingers (or cutting edges) are coming together to make ONE Effective cutting edge.

7.2.1 EXAMPLE 1: THE FACEMILL

Correctly ascertaining the number of effective cutting edges on a cutting tool is the first step to correctly calculating the Feedrate. As shown by the Facemill example on ***Figure 49***, every insert on the Facemill is positioned exactly at the same location axially and radially. Therefore, this Facemill has 8 Effective cutting edges.

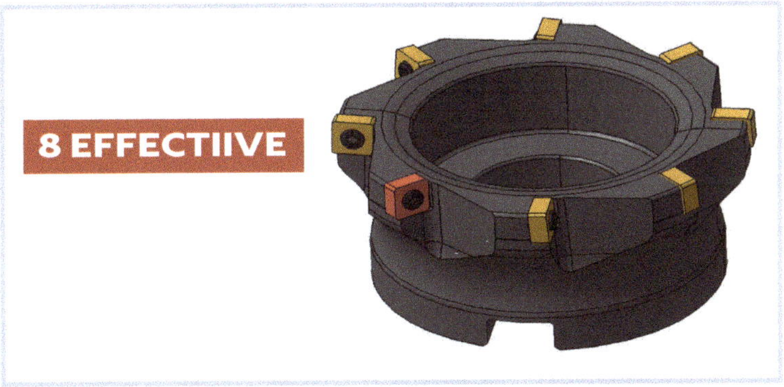

Figure 49. Number of Effective (The Facemill Example).

7.2.2 EXAMPLE 2: THE ENDMILL

Looking at the Endmill example in ***Figure 50***, lets first focus on a single row of inserts as shown highlighted in orange. Each of these inserts has sufficient overlap with the one above or below it so as to create one continuous edge down the helix of the cutter. Additionally, we have two more rows of inserts doing exactly the same thing. So, this Endmill has 3 Effective cutting edges.

Figure 50. Number of Effective (The Endmill Example).

7.2.3 EXAMPLE 3: THE SLOTTING WHEEL SHOWING 9 EFFECTIVE

The slotting wheel shown in ***Figure 51***, is probably one of the best way to gain clarity on number of effectives, as it is most analogous to the 'child scoping a trench in sand' example. As shown highlighted in orange, the inert on the top side of the wheel overlaps with the consecutive insert on the bottom side of the wheel to cover the span of the slot it is trying to create. These two highlighted inserts together form one effective cutting edge. Altogether there are 9 such pairs of inserts, so this slotting wheel is considered to have 9 Effective cutting edges.

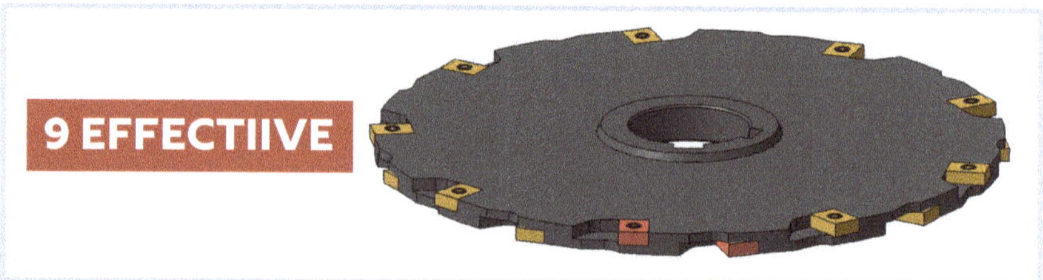

Figure 51. Number of Effective (The Slotting Wheel Example Showing Half Effective).

This is a very good time to bring to light the unique approach of looking at number of effective in drills. As seen in ***Figure 52***, a drill can have numerous flutes that aid in chip evacuation while drilling a hole. However, it would be a very big mistake to look at these flutes and count them up as effectives. The big differentiation of a drill is that it engages and cuts with its central tip, and the cutting edges radiating out to the max diameter. At the very tip of the drill it meets up to a single central point. Therefore, any drill is always going to be One Effective. Deviating from this mental best-practice of considering all drill to be One Effective, can cause very serious errors in the calculations of its Feedrates. This clarity is further discussed in the section on Feedrate calculations for drills with special emphasis on '*Advance per revolution*'.

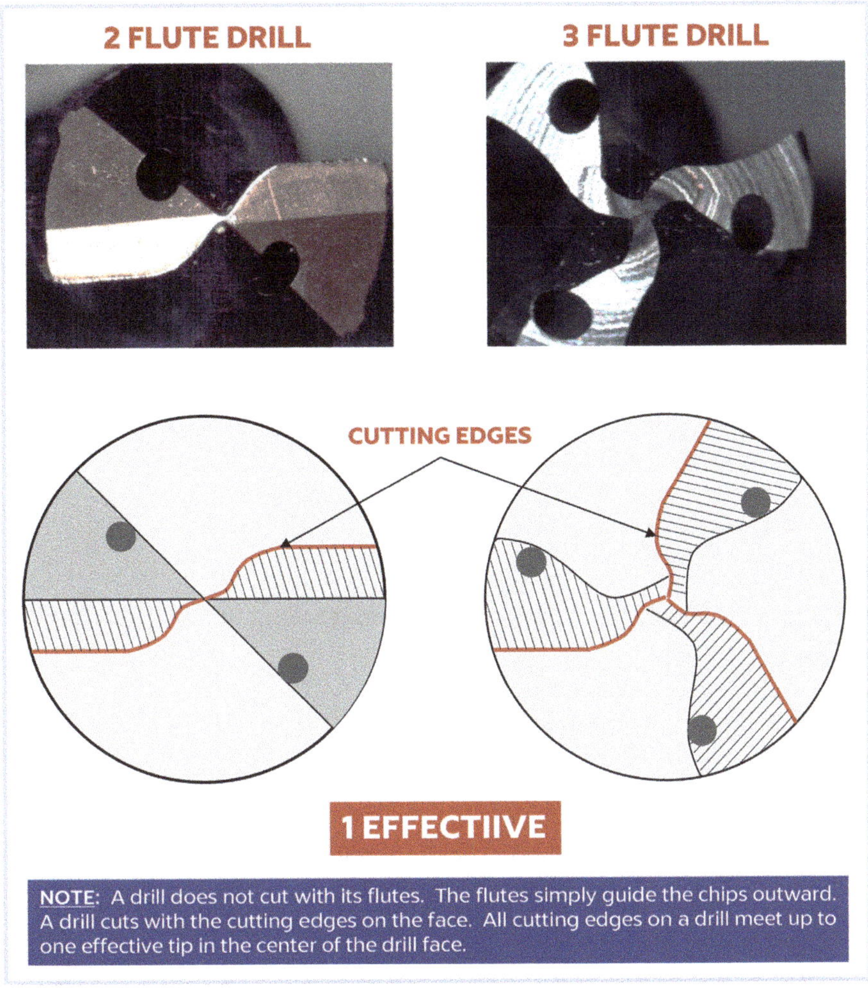

Figure 52. Number of Effective on a Typical Drill.

8

LEAD & RAKE ANGLES

The Lead and Rake angles are the very fundamental descriptors of a cutting edge's geometry. These collective angles serve 3 primary functions. First, they help optimize the efficiency of a cut, thus drawing less power from the machine. Secondly, these angles can help us direct the cutting force vectors that facilitate optimal cutter stability, reduced vibrations, and increase cutter strength. Finally, the lead and rake angels help not only facilitate the curl of the chip, but also guide its direction of its movement. At this point it should be recognized that the discussion on Lead and Rake angles is kept at a basic level of detail to help understand the usage of a cutting tool, rather than teach the details of creating a cutting edge geometry.

8.1 LEAD ANGLES

Let's visualize some everyday examples where we are inherently utilizing the benefits of lead angles, and may not be realizing it. ***Figure 53*** shows some examples of: Scooping ice-cream, shoveling snow, spreading butter on toast, and using a trowel to spread mud compound for tiling. In each of these cases we orient our tool (scoop, shovel, knife, & trowel) in a manner that best facilitates efficiency and functionality. Consequently, creating the least amount of strain on the machinery of our body. Visualize orienting each of these tools in alternate directions while trying to accomplish the same task. Can the job still be done? Possibly, but it would be quite inefficient, wasteful, and not to mention a strain on the body. The concept of Lead Angles in cutting tools works in much of the same way.

As shown in ***Figure 54*** the cutting insert is oriented at about a 45° Lead Angle. When this cutter gets in the cut, it will be subjected to a cumulative Cutting Force (F_T). This Cutting Force could be broken up into its vector components in the Axial direction ($F_{T\,Axial}$), and the Radial direction ($F_{T\,Radial}$). Simply put, various Lead Angles help the user proportion these vectors as required by the application.

Let's talk through the graph in ***Figure 54*** on the effects of altering Lead Angles. As we start to the right side of the graph we observe the cutter insert to have 0° Lead Angle. Here the total cutting force F_T is directly oriented in the radial direction only, and thus carrying only a radial force vector. If the cutting tool is not stiff enough, this purely radial force creates the worst scenario for inducing vibrations and bending.

Figure 53. Some Everyday Examples of Lead Angles.

Next, as the insert is tipped till it gets to the 45° Lead Angle, the predominant previous radial force vector now gets reduced and split into an axial vector component. This is a classic and very effective orientation scenario used in most large Face-milling operations. The machine, holder, and tool system is significantly stiffer in the axial direction, therefore capable of absorbing a lot more forces in this direction, than the radial direction.

As we carry on further to the left on the graph in ***Figure 54***, the Lead Angle now turns to 90°. Here all the cutting forces are directed in the axial direction, and none in the Radial direction. The scenario is a good representation of plunging tools, flat bottom drills, and some high velocity tools (i.e. very shallow axial engagement, but very high speeds).

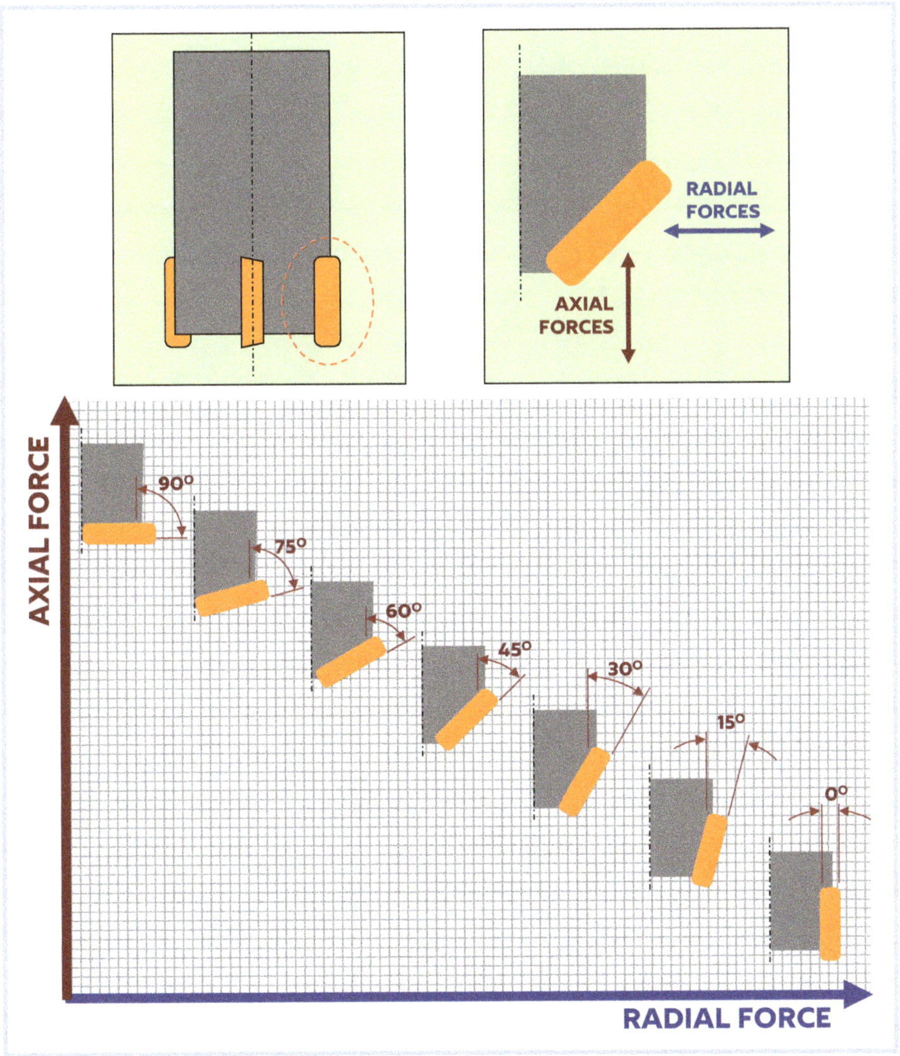

Figure 54. Impact of Changing Lead Angles on Axial & Radial Forces.

8.2 RAKE ANLES

Rake angles can broadly be broken up as Axial & Radial Rakes.

8.2.1 AXIAL RAKE

A depiction of Axial Rake is shown in ***Figure 55***. This is the angle created by the cutting edge with respect to the central axis of the cutting tool. The main functions of Axial Rake Angles are:

- To impact the efficiency of the cut.
- To affect the direction of the cutting forces.

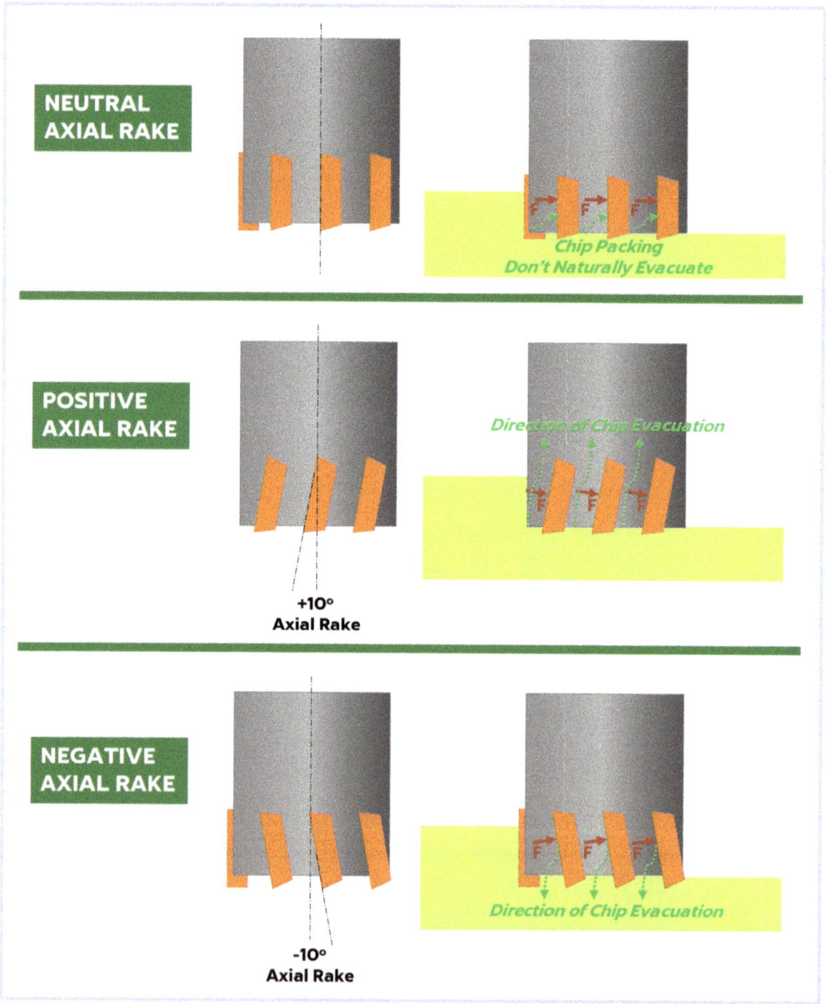

Figure 55. Axial Rake Angles.

The Top illustration in ***Figure 55*** show a Neutral Axial Rake, while the middle and bottom illustrations show a Positive, and Negative Axial Rakes respectively. The Neutral Axial Rake is shown as a visual aid only. Cutters with Neutral Axial Rakes are rarely encountered in industry.

8.2.1.1 POSITIVE AXIAL RAKE

A Positive Axial Rake is shown in the middle illustration of ***Figure 55***. The Positive Axial Rake when in the cut creates forces that tend to PULL on the cutter (vis. Into the workpiece). Additionally, the chips created by a Positive Axial Rake would move upward along the Rake Face of the cutter.

8.2.1.2 NEGATIVE AXIAL RAKE

A Negative Axial Rake is shown in the bottom illustration of ***Figure 55***. The Negative Axial Rake when in the cut creates forces that tend to PUSH on the cutter (vis. Away from the workpiece). Additionally, the chips created by a Negative Axial Rake would move downward along the Rake Face of the cutter.

8.2.2 RADIAL RAKE

A depiction of Radial Rake is shown in ***Figure 56***. While looking down the face to the cutting tool, this is the angle created by the cutting edge with respect to the radius of the cutter. The main functions of Radial Rake Angles are:

- To impact the efficiency of the cut.
- To affect the direction of the cutting forces.

The Top illustration in ***Figure 56*** show a Neutral Radial Rake, while the middle and bottom illustrations show a Positive, and Negative Radial Rakes respectively. The Neutral Radial Rake is shown as a visual aid only. Cutters with Neutral Radial Rakes are rarely encountered in industry.

8.2.2.1 POSITIVE RADIAL RAKE

A Positive Radial Rake is shown in the middle illustration of ***Figure 55***. The Positive Radial Rake when in the cut creates forces that tend to PULL on the cutter (vis. Into the workpiece). Many in the industry like to visualize the Positive Radial Rake as a hook. When dealing with Radial Rakes (and more so with Positive Radial Rakes) special attention needs to be given the clearance on the Flank Face of the cutter. If sufficient clearance in not given to the Flank Face, it would rub against the workpiece prior to the cutting edge ever making contact.

8.2.2.2 NEGATIVE RADIAL RAKE

A Negative Radial Rake is shown in the bottom illustration of ***Figure 55***. The Negative Radial Rake when in the cut creates forces that tend to PUSH on the cutter (vis. Away from the workpiece). The natural tipping of the Negative Radial Rake tends to create an inherent clearance angle in the Flank face.

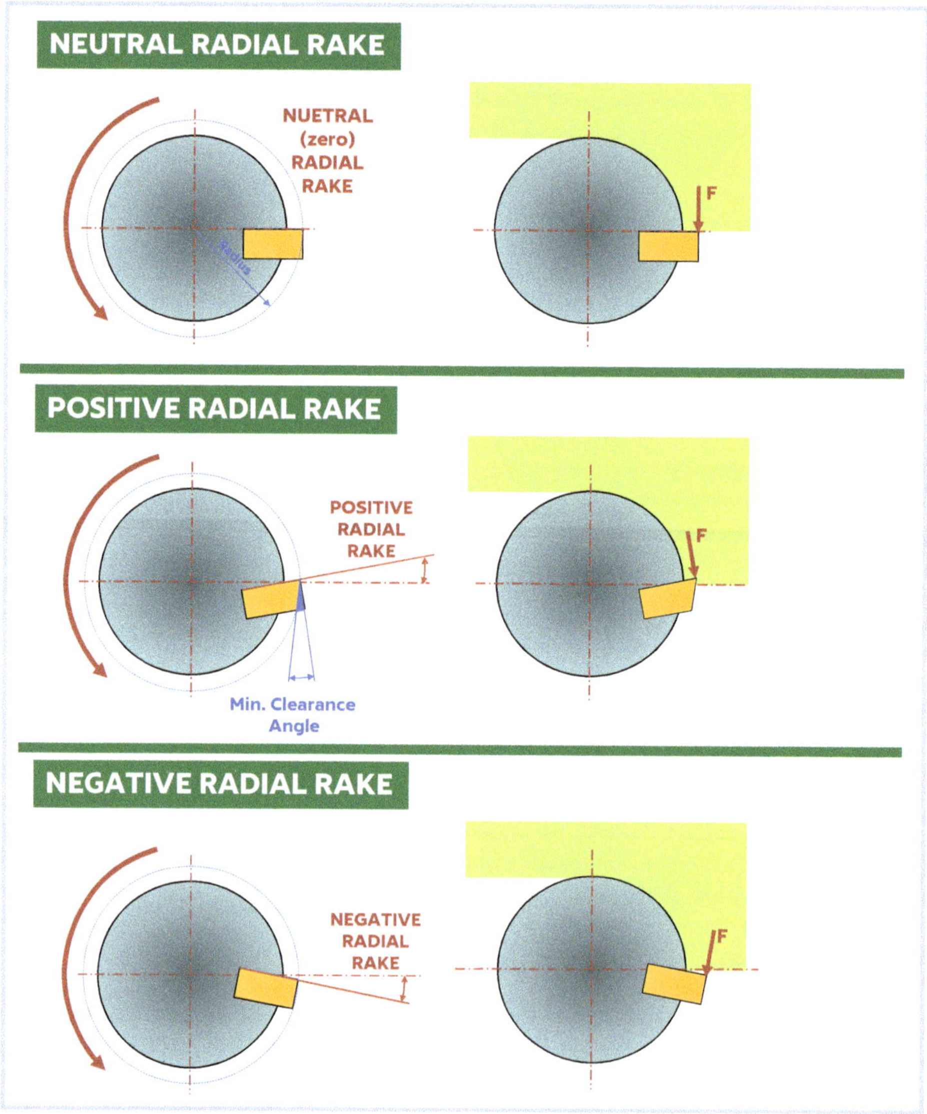

Figure 56. Radial Rake Angles.

8.3 PUTTING IT ALL TOGETHER (LEAD ANGLE, AXIAL & RADIAL RAKES)

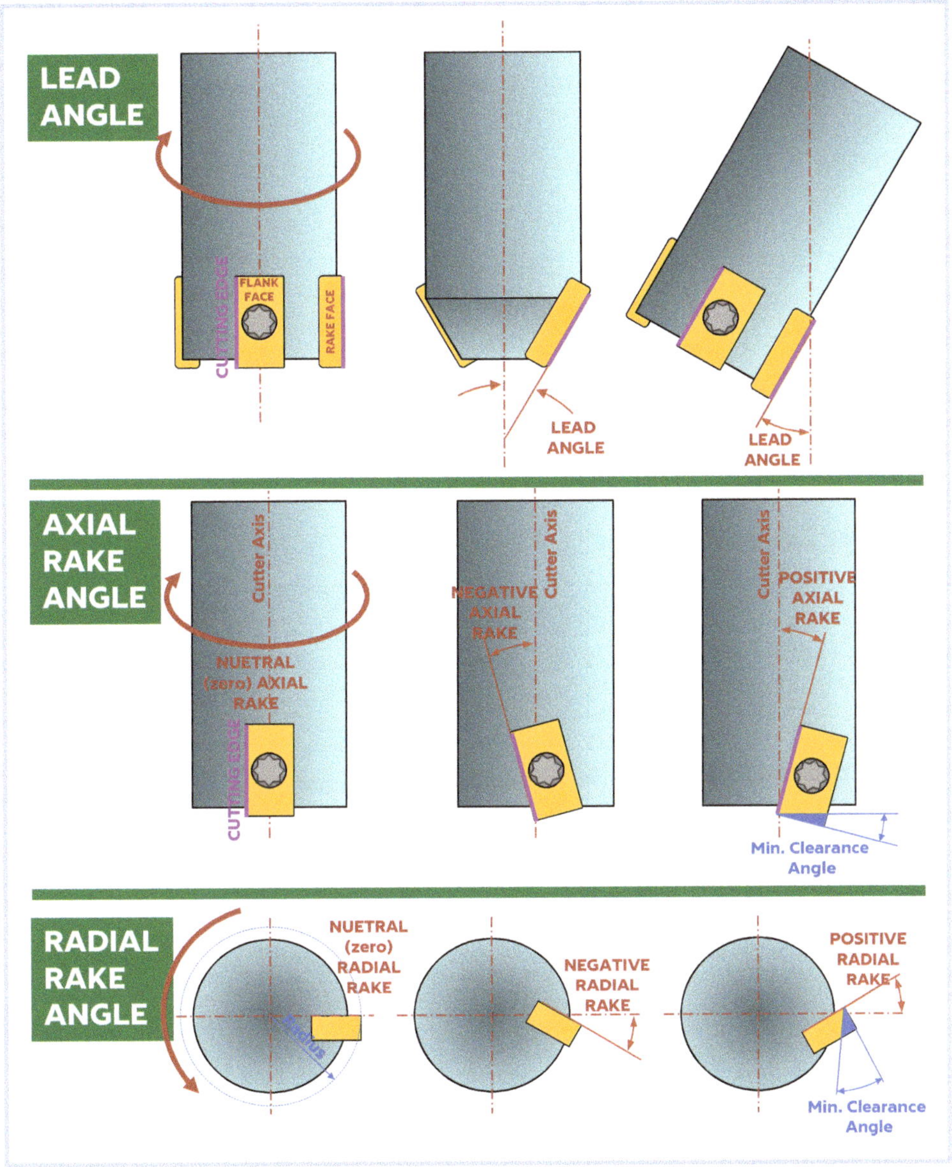

Figure 57. Lead Angles, Axial & Radial Rakes, and Clearances

8.4 ALTERNATE METHODS OF IMPACTING RADIAL RAKE

Let's follow along with the example worked out in ***Figure 58***. For the sake of simplicity lets agree that the application in question requires a 10° positive radial rake, and an idea is floated out that such a tool needs to be engineered. One way to accomplish this is to simply take a rectangular styled insert and tip it ahead of the radial centerline by positive 10° (& exercise the necessary cautions for clearance on the Flank Face).

Another method as shown in ***Figure 58***, is to start with an insert who's Rake Face carries a positive 15° chip breaker type geometry. We could start by assembling such an insert along the radial centerline of the cutter body. However, this would create a scenario where we have a positive 15° radial rake, and not the desired positive 10°. Next, this insert could now be tipped back by a negative 5° radial rake, thus resulting in a cumulative positive 10° radial rake on the rake face. The latter method offers three distinct benefits. First, it naturally creates the clearances required on the flank face. Second, it offers the benefits of a chip breaker on the rake face. And third, in a real life scenario it tips the insert screw at a more ergonomic and manageable angle to facilitate ease of removal and installation.

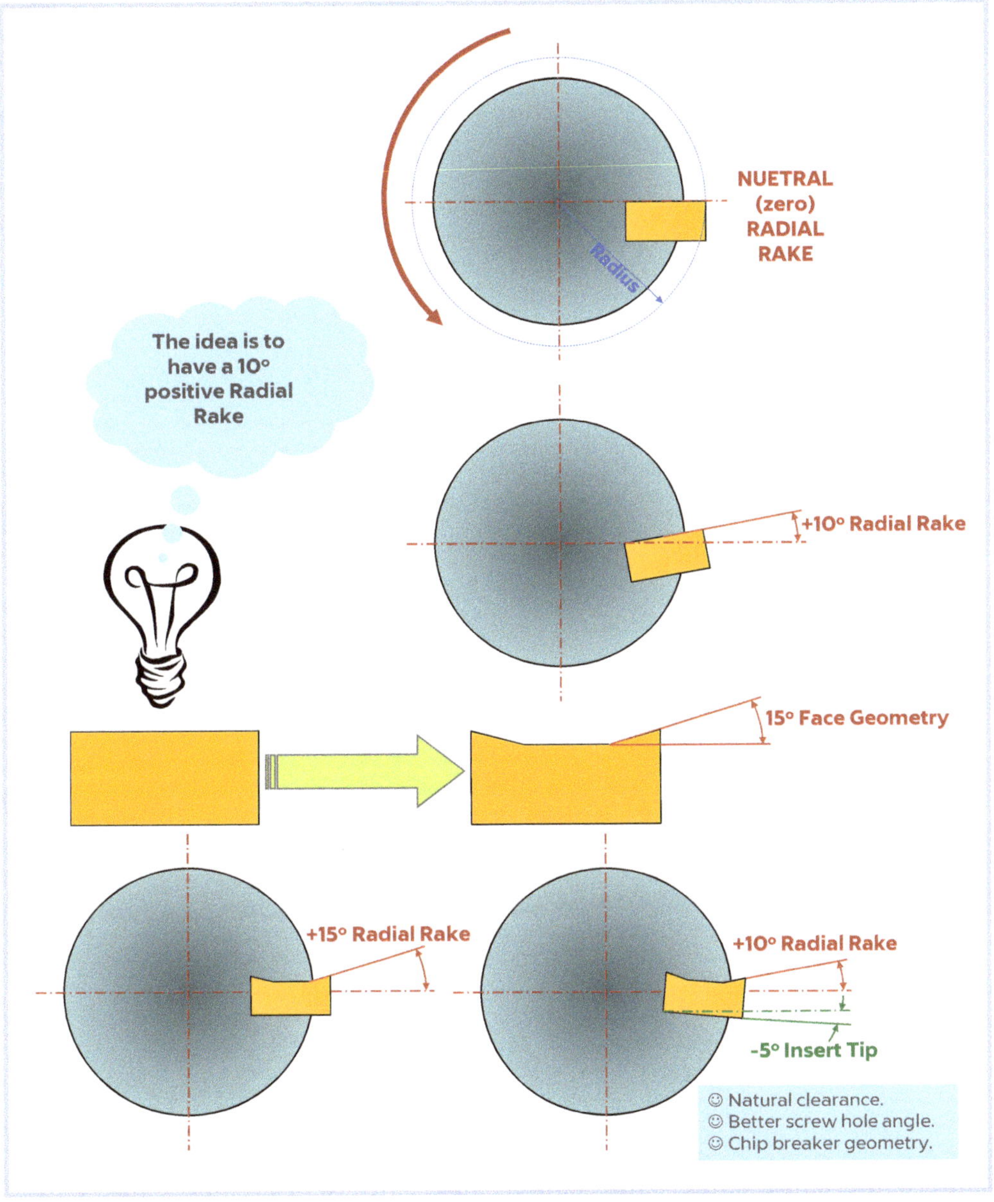

Figure 58. Impacting Radial Rake.

9

EDGE PREPARATIONS

At the simplest level an edge preparation is a treatment given to a theoretical sharp edge in an attempt to improve its performance.

9.1 EVERYDAY EXAMPLES OF EDGE PREPARATIONS

Examples of edge preparations are all around us in everyday objects (***Figure 59***). Compare the edges of a steak knife to that of a butter knife. The steak knife is relatively sharper, while a butter knife is relatively duller. Can we use a butter knife to cut steak? Sure we can, but we may expend more energy and even shake the dinner table. Can we use a steak knife to spread butter on a slice of bread? Sure we can, but we may end up damaging the bread. In either scenario swapping tools creates a less than ideal solution.

Visualize a masonry chisel. This tool carries a significant edge preparation that allows it to stand up to the impact against stone when struck with a hammer. Imagine if this chisel were sharp. Upon use, it would chip off instantly and prove ineffective very quickly.

A surgeon's scalpel is an extremely sharp tool, and although it is used to cut through flesh it is not quite the ideal tool to use at a steak dinner. At the very least, such a behavior would be socially unacceptable. But on a more serious note, the scalpel is engineered to be an instrument of extreme precise cutting. Such a level of precision is not a necessary deliverable while cutting a steak.

Figure 59. Everyday Examples of Edge Preparations.

9.2 FUNCTIONS OF EDGE PREPARATION ON A CUTTING TOOL

The concept of edge preparation has been known to man since the advents made in early blacksmithing. These concepts have carried over in modern day cutting tool technology as well. The big leap in recent year is not the ability to simply put and edge preparation on a cutting edge. Rather, it is the manufacturing ability to put on a very small consistent edge preparation over thousands of cutting edges, to yield a repeatable reliable performance. Significant leaps have also been made in the ability to measure these fine edge preparations through its manufacturing process. ***Figure 60*** shows a typical edge prep treatment given to a sharp edge. There are two primary reasons for applying edge preparations to the cutting edge of a tool:

1. An edge preparation strengthens the edge of the cutting tool. By creating some geometry in place of a sharp edge, we are essentially increasing its

area of contact during cutting. Since stress is defined as force divided by area, as the area of the cutting-edge increases, the stress on that cutting-edge decreases.

CAUTION: The shape and size of an edge preparation needs to be carefully engineered with machine limitations in mind. A larger edge preparation, means larger area of contact with the workpiece. This in turn yields greater friction and consequently greater forces to push through the cut. This results in the machine having to work harder and draw more power. There have been numerous cases where improperly selected large edge preparations have resulted in stalling out of machines.

2 Secondly, an edge preparation can help the user redirect the vectors of the cutting forces seen by the cutting tool. Although the area of contact of an edge preparation with the workpiece is much smaller, it acts in much the same way of redirecting forces as lead and rake angles do to the rake face.

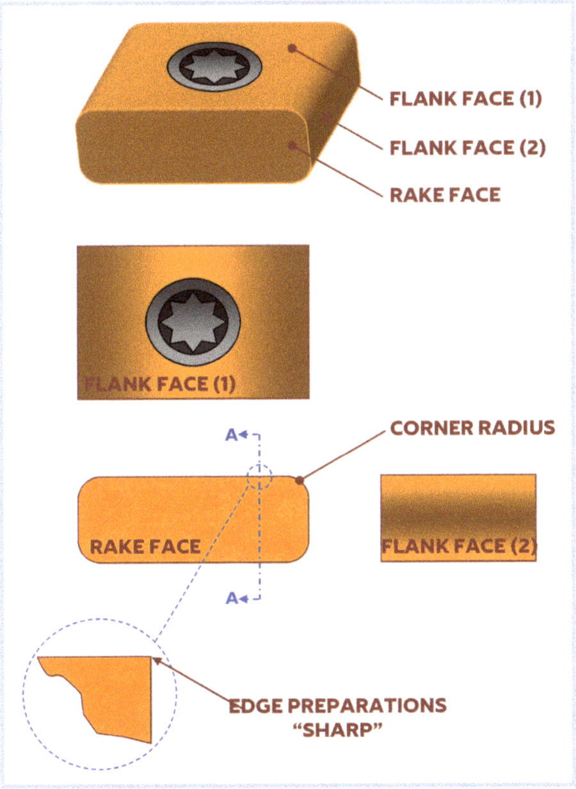

Figure 60. The Cross Section of an Edge Preparation.

9.3 TYPES OF EDGE PREPARATIONS

Edge preparation are broadly broken up a hones (rounding), Lands (chamfers), or a combination of hones and lands.

9.3.1 SHARP

In the world of machining, having a perfectly sharp edge is a very relative term. As a general rule-of-thumb a consistent sharp edge is often used in soft metals such as Aluminum (***Figure 61***). It should be noted that even though it may not be visible to the naked eye, and feels extremely sharp to the touch, some the sharpest of edges carry some version of edge preparation on them for strength.

9.3.2 HONE

One of the most commonly known edge preparations is the hone. Sharp edges are not able to sustain the impact and abuse that comes with machining hard metals (i.e. Steel, Titanium, etc.). To alleviate the stress concentration on a sharp edge, the cutting tool manufacturers create a small rounding on the theoretical sharp edge known as a hone, ***Figure 61***. It is important not to mistake the rounding of a hone with the corner radius of the cutting tool shown in ***Figure 60***. Modern day hones on cutting tools tend to be very small and can usually be appreciated under a microscope. They tend to range from 0.0005" to 0.002". The top illustration in ***Figure 62*** show a magnified example of a hone.

It modern day machining the right hone can make or break an operation. It should be realized that larger the hone, larger the friction component in the cut. In roughing cutters, this frictional component can add up to drawing significantly higher power from the machine spindle. In some cases, improperly selected large hones have been known to stall out a machine. Cutting tool manufacturers are constantly working with their customers to fine tune the hone sizes to best suit the specific application, and maximize productivity.

9.3.3 LAND

Just as the name suggest a land is a flat chamfer like surface created on the theoretical sharp edge, ***Figure 61***. Much like a hone the land serves to spread out the forces and reduce the overall stress on the cutting edge. Some other names for land in the industry are T-Land or K-Land.

The land on a cutting edge is significantly larger than a hone, and can range from 0.002" to 0.010". The middle illustration in ***Figure 62*** shows a magnified example of a Land. Inherent with its benefits are also some drawbacks. This now significantly larger surface in the cut can create a higher power draw on the machine. Additionally, the land can now also act to redirect the cutting force vectors in directions that are desirable. The land adds additional angular geometry to the rake face of a cutting tool, and can play a role in the cumulative rake angle (mostly radial rake) of the cutting tool. A properly engineered land application can stabilize a cut and even eliminate vibrations. A poorly selected land application can catastrophically destroy a cutting tool, and even stall a machine.

9.3.4 COMBINATION OF LAND & HONE

This combination of land and hone is a more realistic scenario in place of just a land (***Figure 61***). Cutting tool manufacturers don't like abrupt changes in geometry where a land meets the rake or flank face. These transitions are often rounded with small hones. Thus, creating a scenario on the cutting edge similar to a hone-land-hone. The bottom illustration in ***Figure 62*** shows a magnified example to a Land & Hone combination.

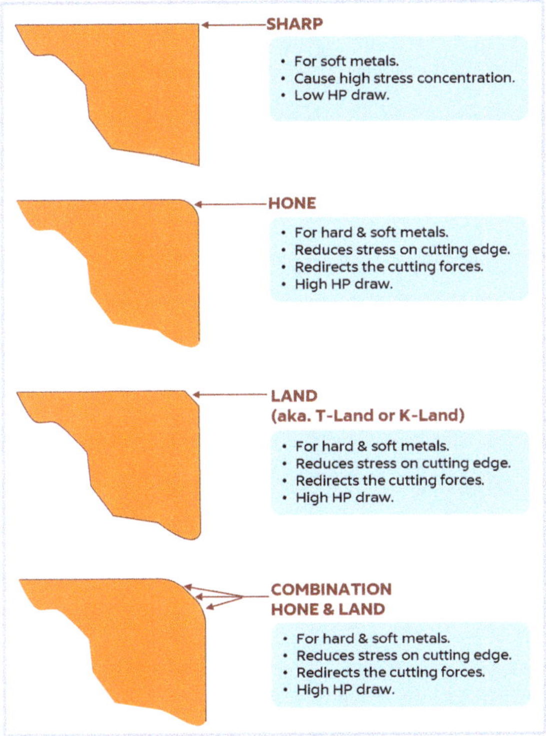

Figure 61. Types of Edge Preparations.

Cutting tool manufacturers can get pretty elaborate and intricate with these combinations of land and hones. A version of this combination is also known in the industry as a waterfall edge preparation.

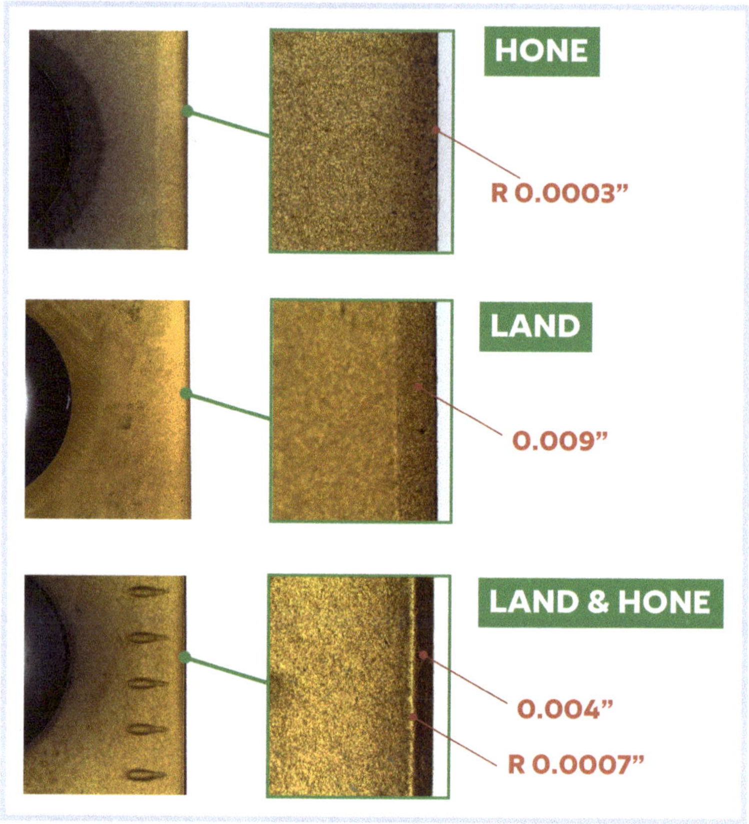

Figure 62. Real Magnified Examples of Edge Preparations.

10

ENGAGEMENT NOMENCLATURE

As in much of this text the bulk of the learning comes from familiarizing the user with the nomenclature of machining. For the sake of simplicity, the abbreviations used in the nomenclature have been kept similar to their verbal descriptions. This method is more aligned with industry nomenclature, rather than research institutes that use the ISO standards (shown in GLOSSARY).

10.1 AXIAL DEPTH OF CUT (DOC)

The Axial Depth of Cut is often just called Depth of Cut (DOC). As seen in ***Figure 63***, this dimension represents how far the cutting tool is engaged axially along its central axis in the workpiece. As from an earlier discussion on 'Cutter Assembly Nomenclature', the DOC should never be larger than the Flute Length.

10.2 RADIAL WIDTH OF CUT (WOC)

The Radial Width of Cut is often just called Width of Cut (WOC). As seen in ***Figure 63***, this dimension represents how far the cutting tool is engaged radially. If the cutting tool as shown in ***Figure 63*** were doing a facing type operation with linear cuts, it would step over radially by a fixed programmed distance and make the next pass. Therefore, the WOC is also known as Stepover.

If the user were not sure on the specifics of how much to radially engage the cutter (WOC), a common starting point **Best Practice** is to use: WOC = 30% of Diameter.

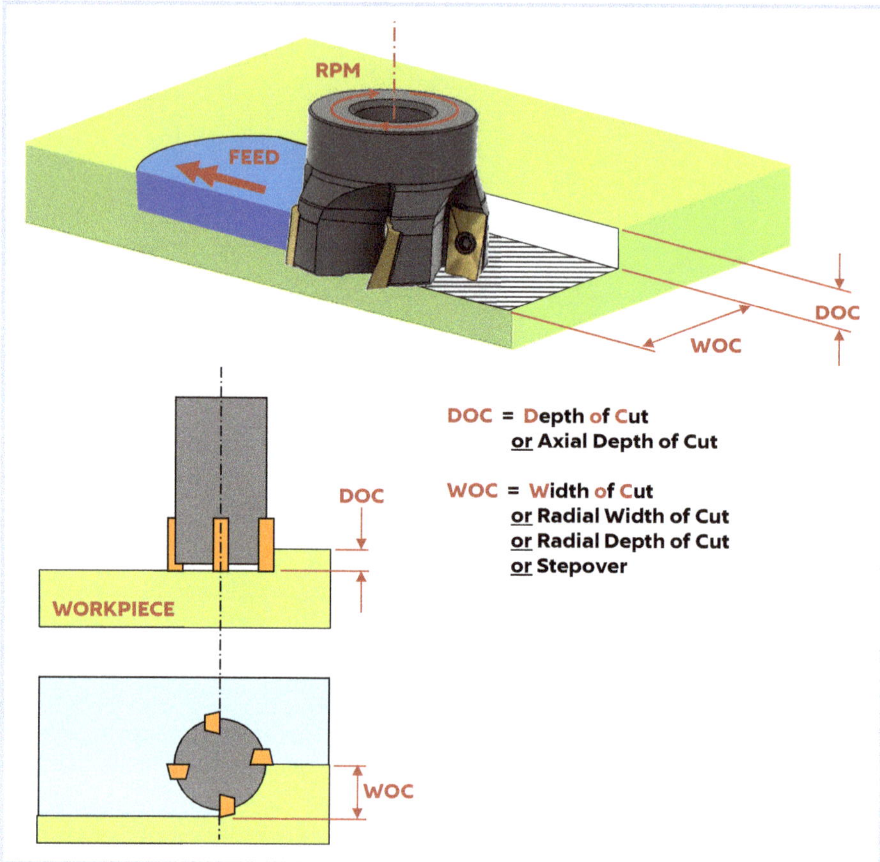

Figure 63. Engagement Nomenclature (DOC & WOC).

10.3 CUTTER PATH APPROXIMATION VS TROCHOIDAL MOVEMENT

At this point it is reasonable to talk about how we represent cutter movements for our future mathematical analysis. Most of the mathematics used in machining related calculations often need to model one singular rotation of the cutting tool.

10.3.1 THE TROCHOIDAL MOVEMENT

The illustration on the left in ***Figure 64***, shows the actual path of a single cutting edge of the cutter, represented by a Start & End point. As this point completes one revolution, it advances in the Feed direction by a pre-determined distance known as Advance Per Revolution (APR). Simply put, in a Trochoidal movement this point is constantly moving forward as it goes around. The illustration on the right in ***Figure 64*** shows a representation of several Trochoidal revolutions stacked together.

10.3.2 COMMON APPROXIMATION OF CUTTER MOVEMENT

In reality a Trochoidal movement is quite complex to model mathematically. To some extent it is almost a combination of an ellipse, and spiral. This degree of modeling complexity is not really necessary for most machining related computation. As seen by the illustration on the right in ***Figure 65***, the Trochoidal movement of a cutting edge can simply be represented by two circles of the cutter's diameter (Start & End), separated by the Advance Per Revolution (APR) distance.

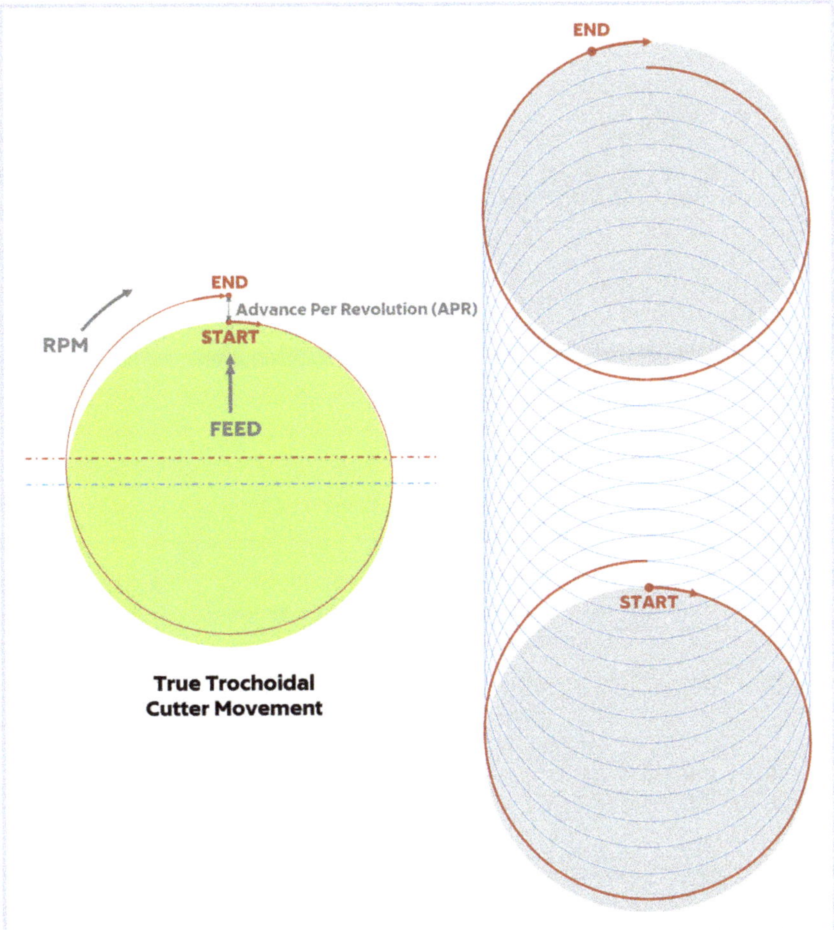

Figure 64. The Trochoidal Movement of a Cutting Edge.

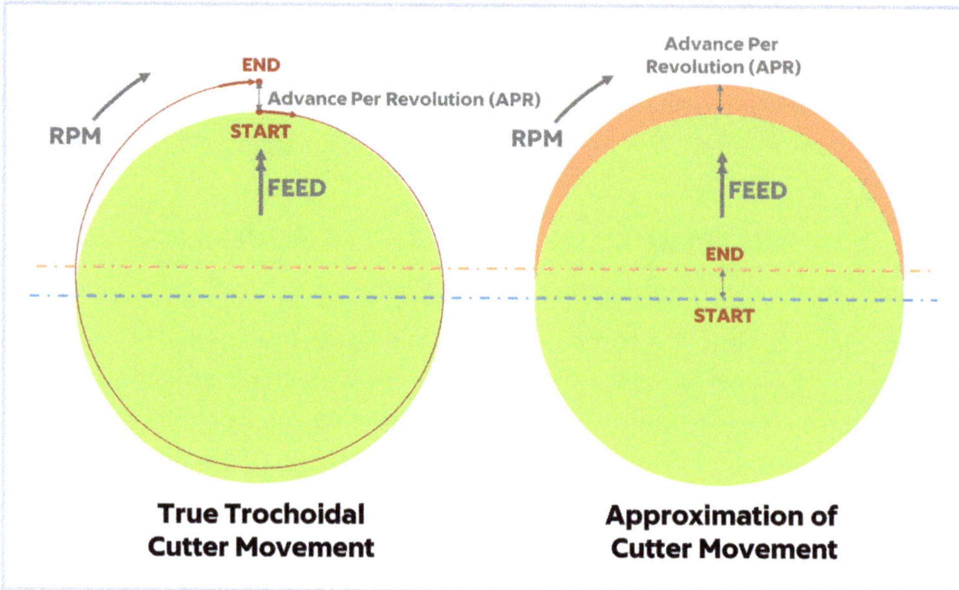

Figure 65. Approximation of Cutter Edge Movement.

11

CLIMB VS CONVENTIONAL MILLING

This concept of Climb vs Conventional milling taps into the kinetics of milling. This topic is of particular importance as there is a distinct preference to the type of cut. The selection is driven primarily by the geometry choices of modern-day cutting tool manufacturers. The easiest way to differentiate Climb from Conventional is to simply look at the images of each (***Figure 66***), and pay special attention to comparing the Chip Thicknesses at the start and end of the cuts. The slightly unique approach taken in this text towards explaining the difference between Climb & Conventional cutting comes from a simple driver; *'to stop bad things from happening'*.

A **Best Practice:** There is one clear and distinctive way to differentiate Climb and Conventional milling: Always look at how the cut starts.

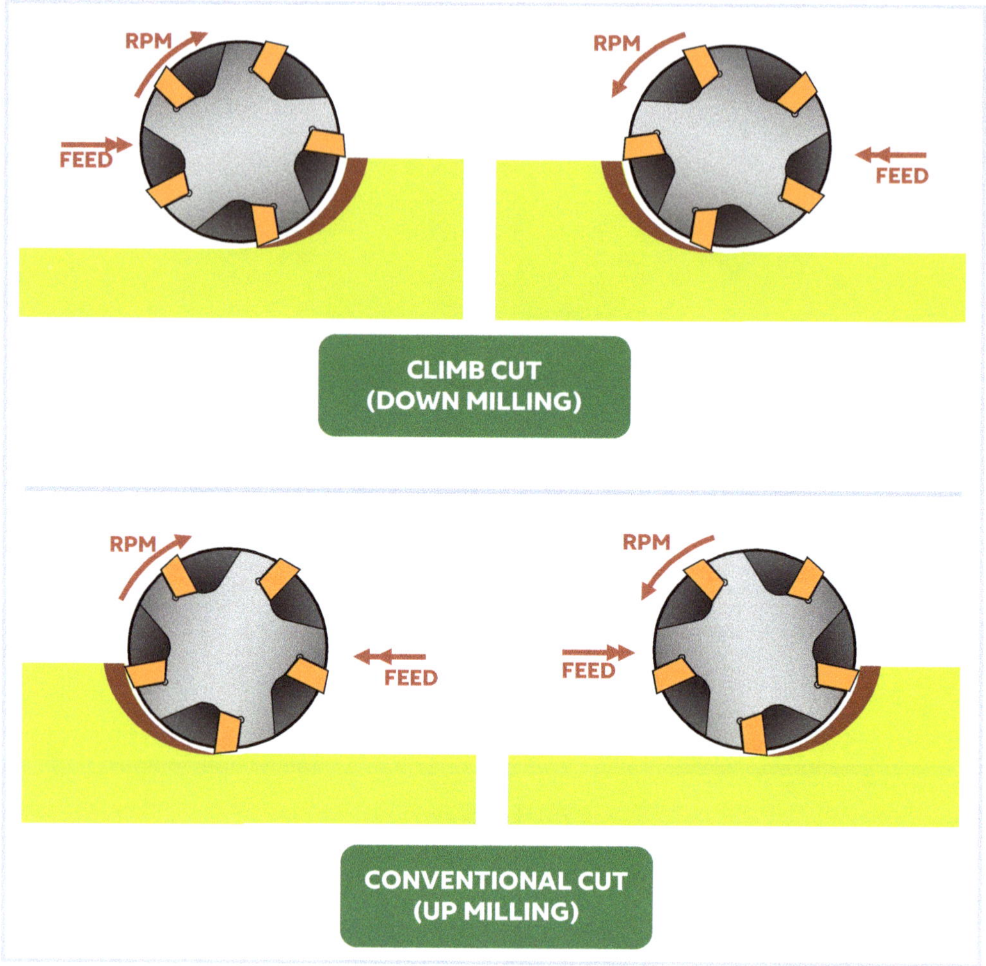

Figure 66. Climb Vs Conventional Milling.

11.1 CLIMB CUTTING

Figure 67 shows a typical Climb milling operation. The Climb cut always starts out with a thick chip. Usually this is the maximum chip thickness. Once the insert is engaged in the cut, the chip thickness then goes from maximum to zero. Some folks like to visualize a car climbing a hill, thus the name. In European countries Climb milling is also known as Down milling. This is because the chip thickness works down from maximum to zero.

Since a climb cut starts with a maximum chip thickness, it initiates with impact to the cutting edge. Therefore, the start of the cut taps into the toughness characteristics of the cutting tool (specifically: the substrate and its Cobalt content).

Next, as the insert sweeps through the cut, it is now confronted by friction and abrasion. Therefore, this part of the cut taps into the hardness characteristics of the cutting tool (specifically the substrate's Carbide content, and Coating heat resistance).

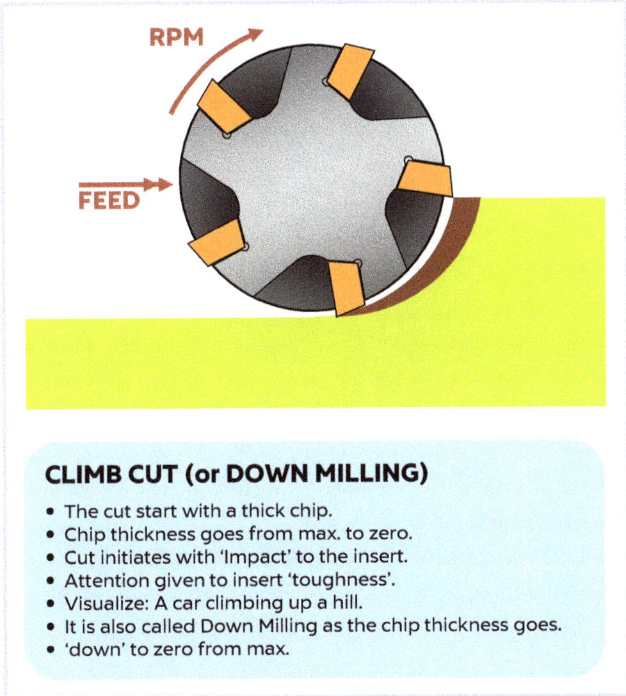

Figure 67. Climb Cut (or Down Milling).

11.2 CONVENTIONAL CUTTING

Figure 68 shows a typical Conventional milling operation. The Conventional cuts always starts with a very thin zero chip thickness. Once the insert is engaged in the cut, the chip thickness then goes from zero to the maximum. Some folks like to visualize Conventional Cutting as earth excavator. In European countries, Conventional milling is also known as Up milling. This is because the chip thickness works up from zero to maximum.

Since a conventional cut starts with zero chip thickness, it initiates with abrasion to the cutting edge. Therefore, the start of the cut taps into the hardness characteristics of the cutting tool (specifically: the substrate's Carbide content). Next as the insert sweeps through the cut, it is now confronted by more friction and abrasion. When the insert comes out of the cut, it almost experiences a negative shock. This is because just prior to coming out of the cut it is engaged in the maximum chip thickness, and then as it exits the cut there is almost a spring-release type effect on it. This release impact although relevant and needs to be accounted for, does not rely as much on the toughness characteristics of the substrate as those required in Climb cutting.

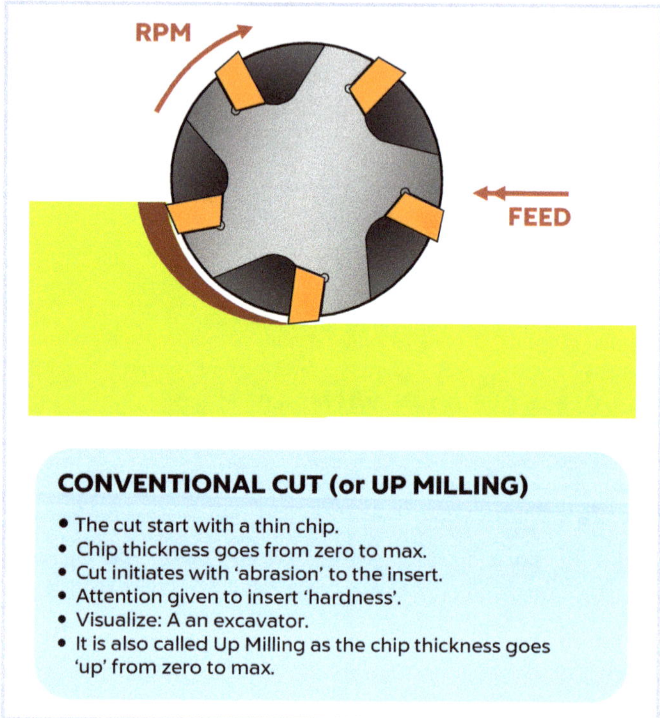

Figure 68. Conventional Cut (or Up Milling).

11.3 A HISTORICAL REFERENCE ON INDUSTRY CHANGES

While it is important to understand the kinematics of climb vs conventional cutting, its full appreciation may be realized by the historical evolution of the two engagement types. It is estimated that prior to mid 1980s cutting tools were predominantly negative in geometry (i.e. negative axial & radial rakes). The machine tools of that era were extremely robust and some may say over-engineered. They had substantially more mass, power, torque and rigidity than that which could be demanded by most cutting tools of that time. The combination of these two factors allowed for conventional cutting to be the primary mode of engagement.

As the industry changed through the 1980s and into the 1990s, there arose the need for energy efficient machines. Decisions started be made based on building just the right machine for just the right operation. Sensitivities to power, torque, and rigidity were heightened. This in turn drove out parts that were with less mass & dampening characteristics. This change collectively drove out the need for freer cutting tool, and consequently led to the positive geometries in cutters. Positive geometry cutters are highly sensitive to the starting conditions of the cut (especially the clearances needed for starting a zero chip thickness cut). The only way to make these highly positive geometries work was to engage them in a climb cutting mode of engagement. This point is most clearly understood by looking at the kinetics of cutting in a Channel Cut in the following section.

11.4 HOW TO LOOK AT A CHANNEL CUT

A Channel cut is when the entire diameter of the cutter is engaged as WOC. Simply ask, is a Channel cut Climb, or Conventional? When this question is asked, there are countless ideas, opinions and perceptions. Once again for the sake of this text we chose to formulate an opinion base on two factors already mentioned:

1. Look at the start of the cut.
2. We want to stop bad things from happening.

As seen in ***Figure 69***, a Channel cut always starts this a thin zero chip thickness. It then works its way up to the maximum chip thickness along the centerline of feed direction of the cutter. Next, the cutting edge works its way back down to a zero chip thickness before exiting the cut. Therefore, following our guidelines on how the cut starts (i.e. zero chip thickness), a Channel Cut is always going to be a Conventional Cut.

Before we go much further it is important to understand the actual path that a cutting edge would trace. As explained earlier in ***Figure 65***, the path of a cutting edge resembles more of the Trochoidal move shown on the illustration on the left-hand side, rather than the commonly used approximation of two circles separated by the APT as shown on the right. Accurately tracing out the Trochoidal path of a cutting edge will help clearly visualize the clearance impacts of Positive, Neutral, and Negative radial rakes during Channel cutting. Now let's see how looking at the start of a channel cut helps us stop bad things from happening.

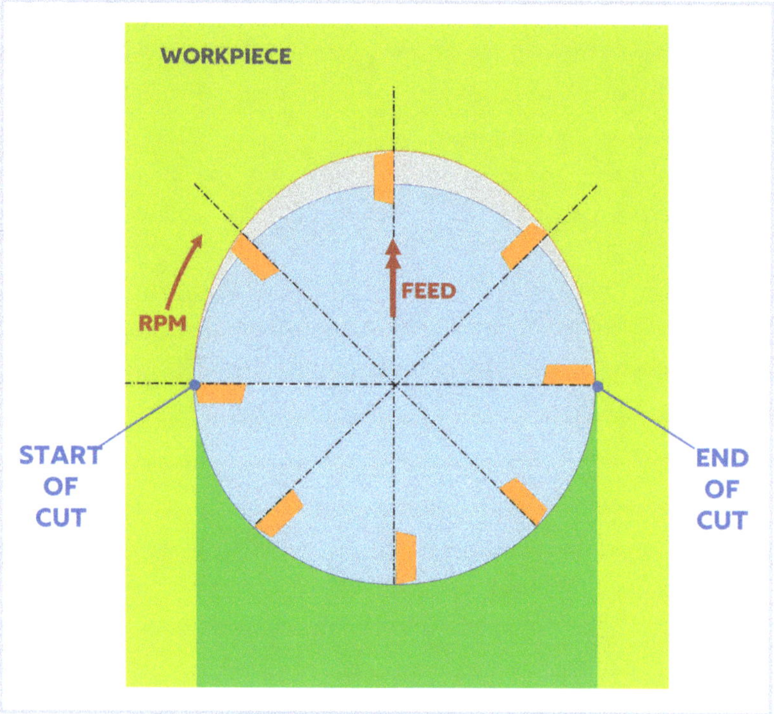

Figure 69. A Channel Cut.

11.4.1 CHANNEL CUT WITH A POSITIVE RADIAL RAKE

As explained, the current cutting tool industry is predominantly creating cutters with positive radial rakes. Now let's see what could happen to a positive radial rake edge geometry as it starts this particular cut at zero chip thickness. ***Figure 70*** shows the progression of a single cutting edge tracing a true Trochoidal path as it travels through a Channel Cut (starting at 0°, all the way through 90°). Our ideal desire is that the cut starts with instantaneous contact of the actual cutting edge to the workpiece. However, since the cutting edge presented to the workpiece is tipped by a positive radial rake, it creates the scenario where the cutting edge is not the first contact with the workpiece due to lack of clearance. Rather, it is the flank face that bangs into the start point of the chip. This is in fact the case (Rubbing on Flank Face) as the cutter then revolves from 0 ° to 60°. The cutting edge only bites into the cut from the 75° to 90 ° section. It is this initial unfavorable abrasion of the flank face that creates the start of many a vibration and consequently premature failure of the cutting tool.

Channel cutting with modern day positive angled cutters (in hard metals) has been the root cause of many inadvertent catastrophic failures. Therefore, by the above deductive reasoning we chose to label a Channel cut to also be a Conventional Cut. If you chose to take this approach; 'bad things won't happen'.

A Best Practice: Occasionally a user may be faced with a scenario where they have no choice and absolutely have to make a Channel Cut with a modern day Positive Radial Rake type cutter. In such a case, it is a best practice to slow down the programmed Feedrate by 50% to make the cut.

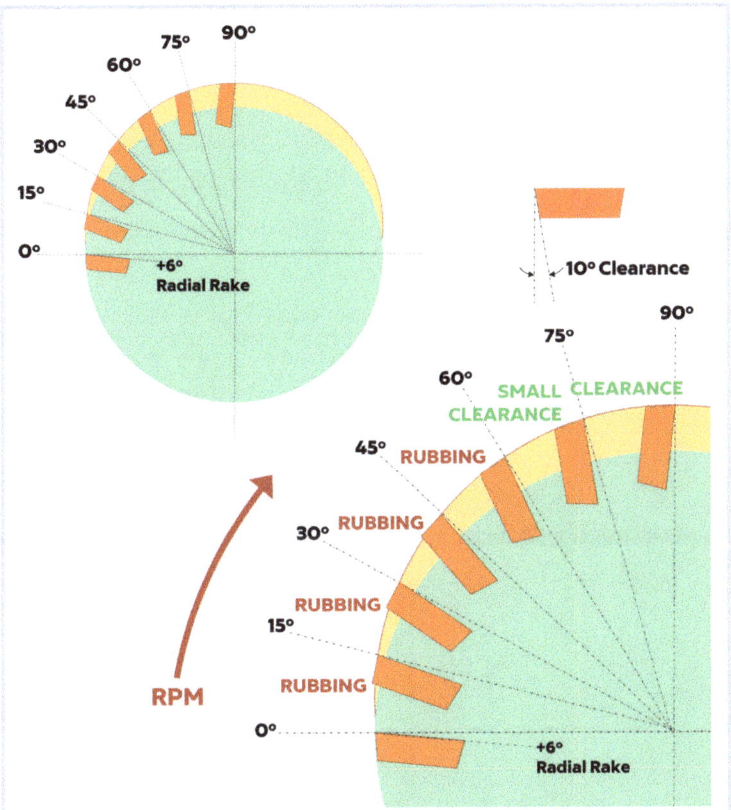

Figure 70. Channel Cutting with a Positive Radial Rake.

11.4.2 CHANNEL CUT WITH A NEUTRAL RADIAL RAKE

Now let's see what could happen to a neutral radial rake edge geometry as it starts this particular cut at zero chip thickness. ***Figure 71*** shows the progression of a single cutting edge tracing a true Trochoidal path as it travels through a Channel Cut (starting at 0°, all the way through 90°). The engagement behavior here is similar to that of a Positive Radial Rake. Our ideal desire is that the cut starts with instantaneous contact of the actual cutting edge to the workpiece. However, since the cutting edge presented to the workpiece has a neutral radial rake, it creates the scenario where the cutting edge is not the first contact with the workpiece due to lack of clearance. Rather, it is the flank face that bangs into the start point of the chip. This is the case (Rubbing on Flank Face) as the cutter then revolves from 0° to 45°. The cutting edge only bites into the cut from the 60° to 90° section. It is this initial unfavorable abrasion of the flank face that creates the start of many a vibration and consequently premature failure of the cutting tool.

Therefore, by the above deductive reasoning we chose to label a Channel cut to also be a Conventional Cut. If you chose to take this approach; 'bad things won't happen'.

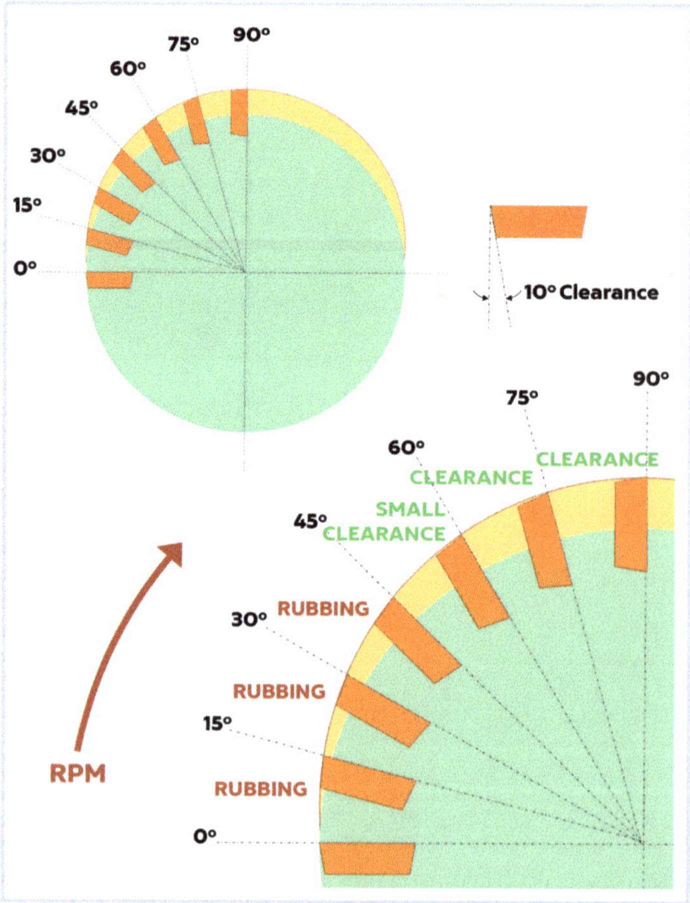

Figure 71. Channel Cutting with a Neutral Radial Rake.

11.4.3 CHANNEL CUT WITH A NEGATIVE RADIAL RAKE

As explained earlier, cutters with a Negative Radial Rakes were common in much earlier cutter generations, and the industry is now trending towards Positive Radial Rake cutter. Now let's see what could happen to a negative radial rake edge geometry as it starts this particular cut at zero chip thickness. ***Figure 72*** shows the progression of a single cutting edge tracing a true Trochoidal path as it travels through a Channel Cut (starting at 0°, all the way through 90°). Our ideal desire is that the cut starts with instantaneous contact of the actual cutting edge to the workpiece. A Negative Radial Rake cutting edge geometry is the only scenario where this actually happens. Unlike the Positive and Neutral Radial Rake scenarios, the Flank Face in a Negative Radial Rake cutter does not bang into the workpiece at the start point of the chip. The cutting edge bites into the cut all the way from the 0° to 90°.

A Negative Radial Rake geometry, although ironically is not a current industry standard, is the only one that is inherently capable of make a successful Channel Cut.

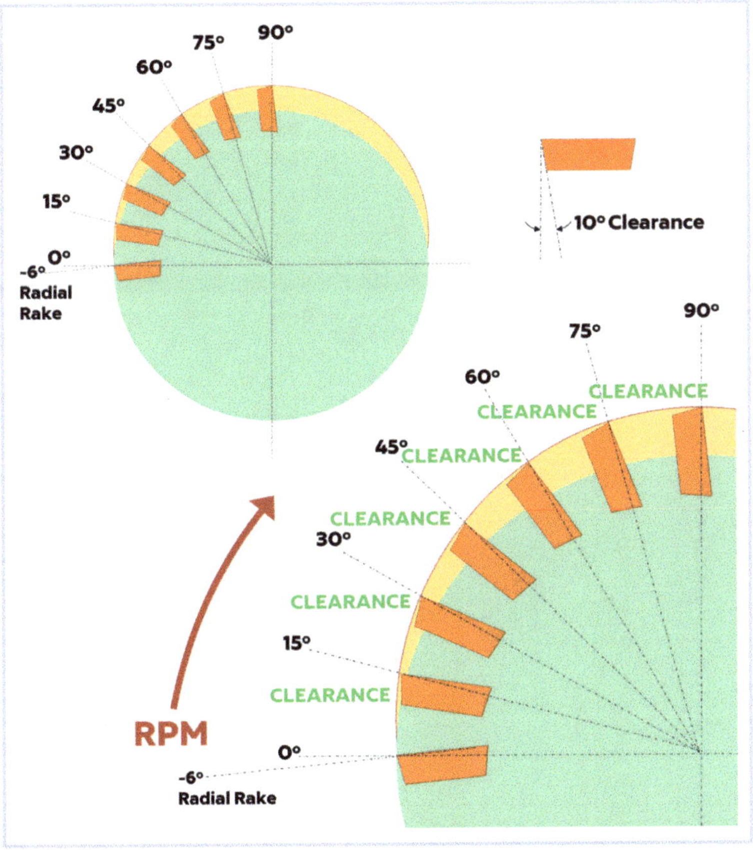

Figure 72. Channel Cutting with a Negative Radial Rake.

12

SPEED & FEED

Speed and Feed are the cornerstone parameters for operating a cutting tool optimally. It is the most fundamental of nomenclature and lingo thrown around by professionals in the machining industry. For the sake of simplicity, the abbreviations used in the nomenclature have been kept similar to their verbal descriptions. For Example: the term 'revolutions per minute' has been abbreviated as 'RPM', rather than 'n' which is the ISO nomenclature. This method is more aligned with industry nomenclature, rather than research institutes that use the ISO standards (shown in GLOSSARY). The main intent of keeping the abbreviations 'user-friendly' is so that the reader can easily visualize the core components of the equations, rather than be forced to memorize them.

Visualizing our lawnmower analogy to milling, the Speed in a cutting tool can be visualized as the rotational speed of the outermost tip of the lawnmower blades. The Speed is actually an abbreviation for Surface Speed, or Surface Feet per Minute (SFM). This is the surface speed of the periphery of the cutter and is mathematically related to Revolutions Per Minute (RPM). The Feed on the other hand is analogous to the velocity of the lawnmower as it is pushed along the grass being cut. The Feed is actually abbreviation for Feedrate, or Inches per Minute (IPM).

12.1 SPEED - SURFACE FEET PER MINUTE (SFM)

The best way to understand and eventually calculate the Surface Feet Per Minute (SFM) is to visualize it as shown in ***Figure 73***. SFM is the speed of any single point on the outer most periphery of most cutting tools, which is usually the cutting edge.

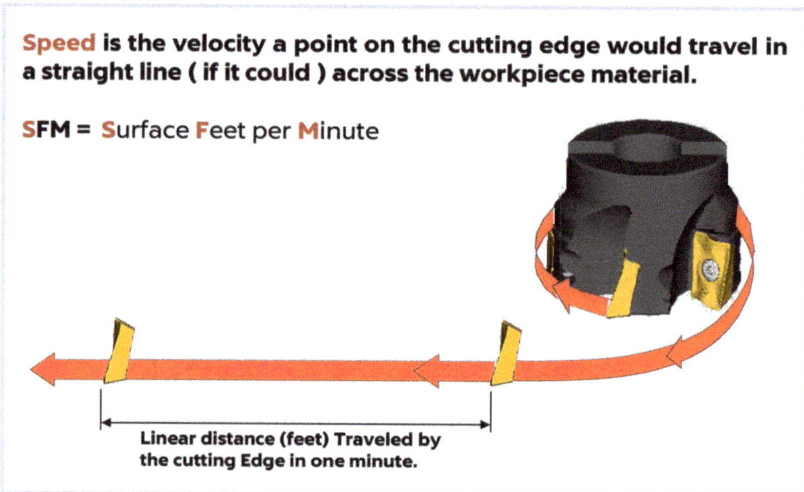

Figure 73. Surface Speed.

The breakdown of the equations for Surface Speed can be understood as follows (***Figure 74***):

- As a single point on the cutting edge goes once around, it covers a single circumference, $\pi \times DIA$. This is the distance covered circumferentially over one revolution.
- Next, we ask ourselves, how many such circumference would be completed in one minute? This is determined by the Revolutions Per Minute (RPM). We now multiply the circumference by RPM. Consequently, we get $(\pi \times DIA).RPM$.
- Historically the SFM is noted in units of feet, and since the diameter of the cutter is in inches, we need to do one final unit conversion of dividing by 12 in/ft. Therefore:

$$SFM = \frac{(\pi \times DIA).RPM}{12} \qquad [Units : ft / min]$$

As discussed earlier in **Chapter 3**, the SFM is one of the dominant control factors for heat in the cut, and consequently related to the hardness characteristics of the cutting tool. If the SFM is higher than what the material of the cutting tool is designed to accommodate, it will result in higher than normal heat in the cut, and excessive abrasive wear of the cutting tool. This will eventually lead to premature, or even catastrophic failure of the cutting tool.

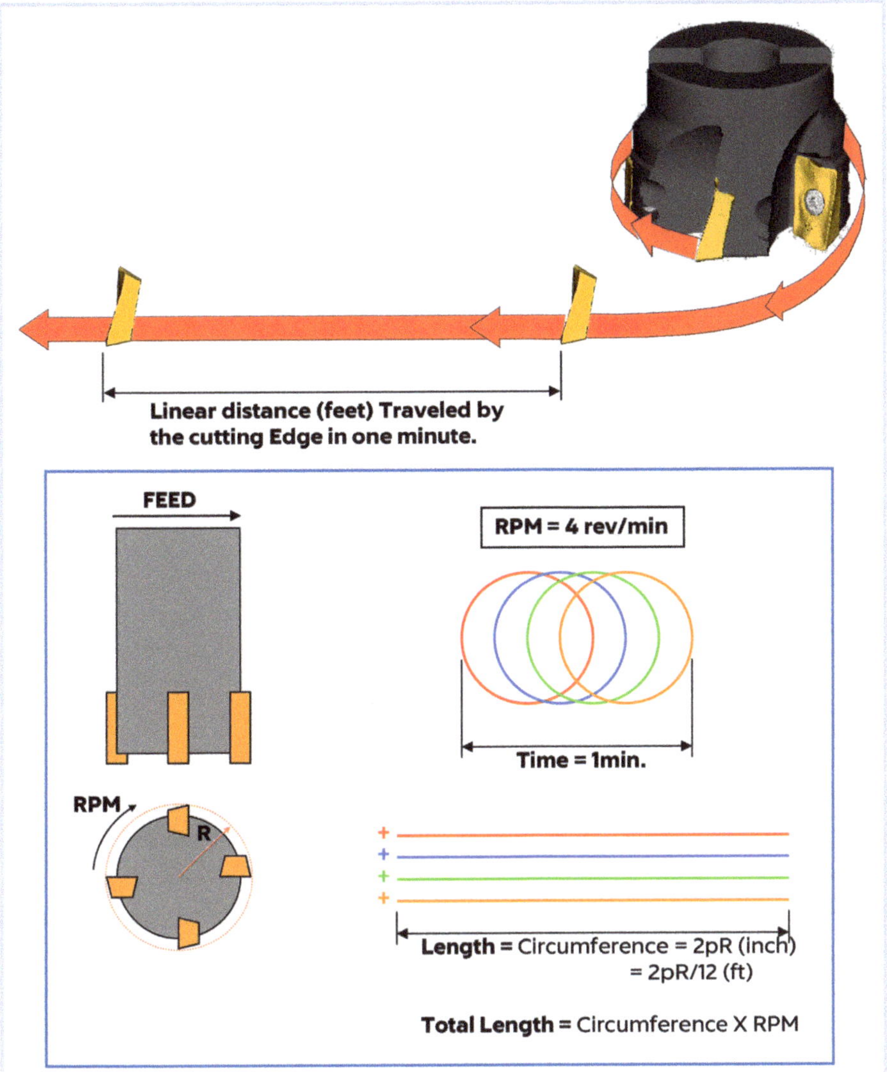

$$SFM = \frac{(\pi D) \times RPM}{12}$$

Figure 74. Calculating Surface Feet per Minute (SFM).

12.1.1 RELATING REVOLUTIONS PER MINUTE (RPM) TO SURFACE FEET PER MINUTE (SFM)

As is quite evident by its name the Revolutions Per Minute (RPM), is directly related to SFM (***Figure 75***).

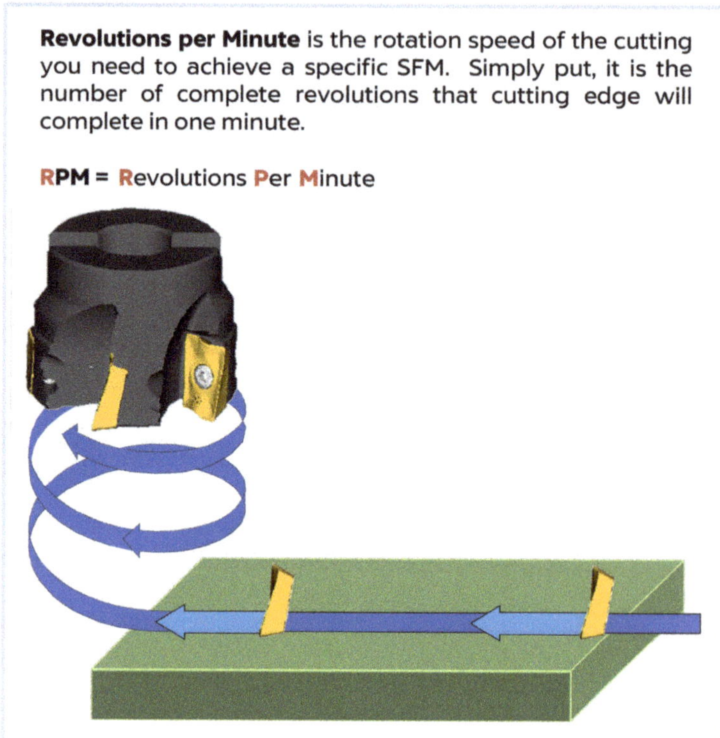

Figure 75. Revolutions Per Minute (RPM).

With some basic algebraic manipulation, the RPM can be calculated as (***Figure 76***):

$$RPM = \frac{SFM \times 12}{(\pi \times DIA)} \qquad [Units : revs / min]$$

It is important to note that even though the SFM is one of the main control factors for heat in the cut, as well as mathematical driver for RPM calculations, it is almost never seen numerically in an NC program or at a machine's control panel. It is RPMs that show up in the NC code outputs, and often read on machine's controller as well.

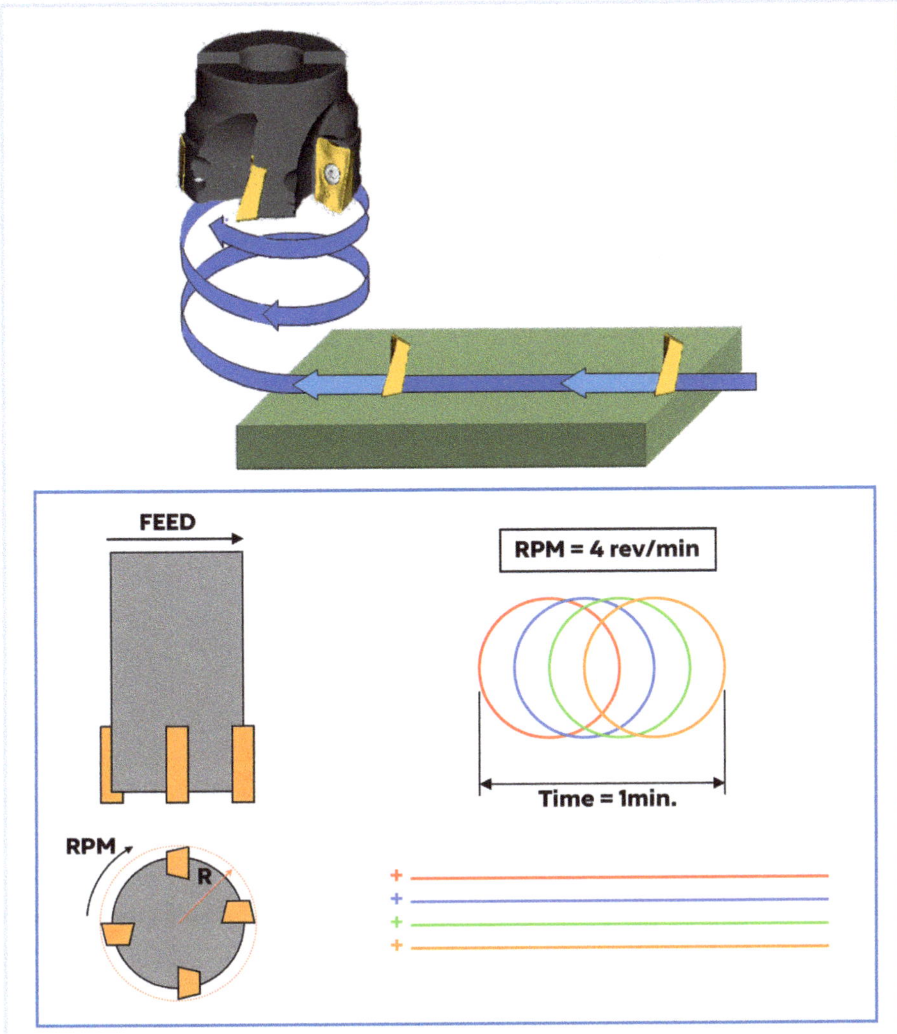

$$RPM = \frac{SFM \times 12}{(\pi D)}$$

Figure 76. Calculating Revolutions Per Minute (RPM).

12.1.2 LOCK THE SFM & CALCULATE THE RPM

To close out the discussion on Speed, let's observe ***Figure 77***. Since Surface Speed is the speed of a cutting edge on the periphery of the cutter, it has the most direct correlation to friction, and heat. Therefore, higher the Speed, higher the Friction and consequently heat in the cut. When focusing at Surface Speed, one needs keep an eye on the abrasive wear of the cutting edge. Consequently, hardness is the material property of the cutting tool that resists abrasive wear, and needs to be kept a watchful eye on.

A Best Practice: Surface Speed is predominantly responsible for the heat in the cut. For each workpiece material we want to make sure that this heat signature remains about the same throughout the spectrum of cutters that are collectively responsible for machining the part. This family of cutters can range in diameters. Therefore, it is important to lock in the Surface Speed for a given workpiece material and calculate the RPMs (***Figure 77***).

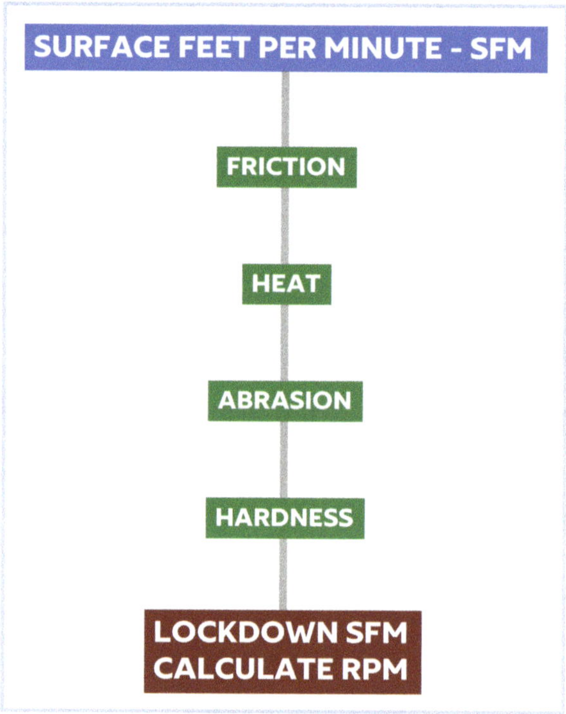

Figure 77. Surface Speed and its Variables.

The effects of calculating RPMs for a fixed SFMs on cutter diameters can be graphically seen in ***Figure 78***. In order to maintain a constant SFM (i.e. heat signature on the periphery of the cutter), the RPM must be slowed down as the cutter diameter increases.

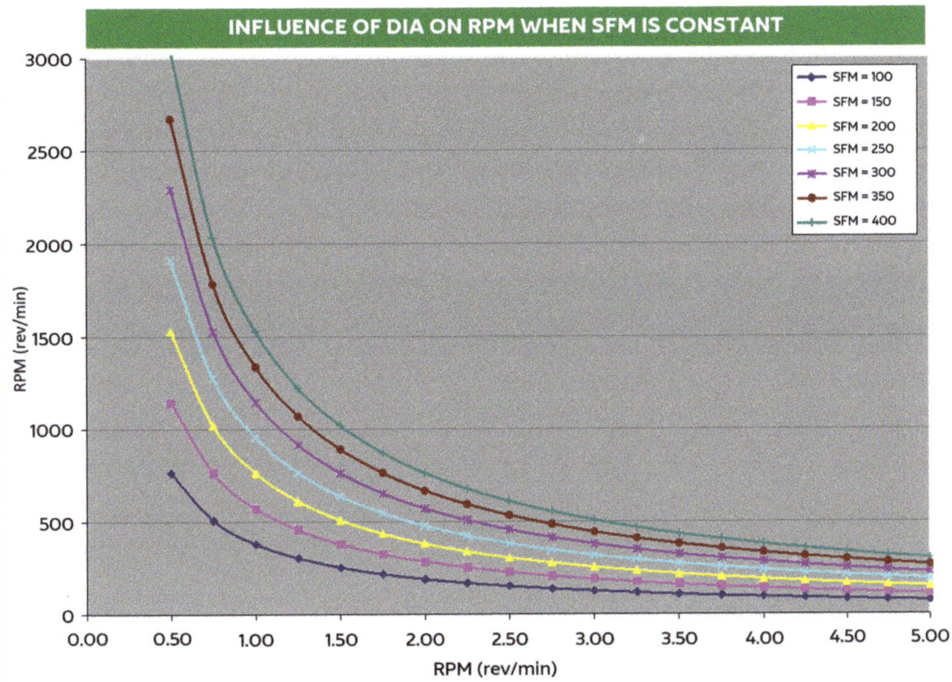

- For a fixed SFM: as DIA increases, RPM decreases.
- For a fixed DIA: as SFM increases, RPM increases.
- At the larger DIAs, a small change in RPM will greatly impact on SFM.
- SFM impacts heat in the cut.
- Always lock-in SFM for a material, and calculate the RPM at the required DIA.

Figure 78. The Effects of Calculating RPMs for a Fixed SFM on Varying Cutter Diameters.

As a starting reference point ***Table 3*** shows some good starting point reference SFM values for Roughing & Finishing cuts in some commonly machined work-piece materials. Simply put these speeds are the baseline status quo operating parameters for modern day coated carbide cutter so that they yield a repeatable reliable performance and tool wear. They are arrived at empirically by way of testing, and real-life production experience.

MATERIALS	SFM – Roughing (ft/min)	SFM – Finishing (ft/min)	WET / DRY
Aluminum 7050	10,000	12,000	WET
Steel 4340	500	900	WET OR DRY
Stainless Steel	300	500	WET OR DRY
Stainless (PH)	400	600	WET OR DRY
Iron	800	1000	DRY
Titanium 6Al4V	140	200	WET
Titanium 5553	65	65	WET
Inconel	100	120	WET
Magnesium	10,000	12,000	WET
C465-H1000	140	150	WET
C465-H950	130	150	WET

NOTE: Data is based on Carbide cutting tools.

Table 3. Some SFM Values for Commonly Machined Workpiece Materials.

The example shown in ***Figure 79*** walks us through step-by-step scenario of calculating RPM for a given SFM in a known workpiece material. Also shown in this scenario is the impact of changing cutter diameters for a fixed SFM. To maintain a fixed SFM for a given workpiece material, as the cutter diameter increases, the RPM decreases. Also, shown in this scenario is the detrimental impact of not adjusting the RPMs for changing cutter diameters.

- We are rough machining Titanium 6Al4V, using a 2" DIA Indexable endmill. A NC programmer wants to know what RPM to run the cutter at?

- Start by locking in a SFM of 140 ft/min.
- Select the correct formula.

$$SFM = \frac{(\pi D) \times RPM}{12} \quad \Rightarrow \quad RPM = \frac{SFM \times 12}{(\pi D)}$$

$$RPM = \frac{140 \times 12}{(\pi 2)} = 267 rev/min$$

- We decided to change from a 2" DIA to a 3" DIA tool. What is the new RPM?

$$RPM = \frac{SFM \times 12}{(\pi D)}$$

$$RPM = \frac{140 \times 12}{(\pi 3)} = 178 rev/min$$

- What would have happened if we did not lock in SFM, but rather have used the 267 RPM in the 3" DIA tool?

- Lets check the SFM at 267 RPM for a 3" DIA tool.

$$SFM = \frac{(\pi D) \times RPM}{12}$$

$$SFM = \frac{(\pi 3) \times 12}{12} = 210 ft/min$$

- 210 SFM would be too high for Ti 6Al4V.
- Cutter would get too hot and prematurely fail.

Figure 79. Example 1 of SFM Calculations.

The example shown in ***Figure 80*** walks you through a step-by-step scenario of troubleshooting a typical damaged cutter scenario. It also coaches you through how to make a recommendation to remedy the problem by using reference data from ***Table 3***.

- An operator has just called on you for some support in regards to a suspect Ti 6Al4V roughing operation. When you meet the operator at the machine you learn that the tool in concern is a 2" DIA indexable endmill. The machine controller indicates that it is running at 573 RPM. The complain from the operator is that the tool which previously ran for 90 minutes, now only lasts 20 minutes. The operators suspects the quality of the cutting tool. How should you trouble shoot this operation?

- Recognize the known: DIA = 2", RPM = 573.
- Recognize premature tool failure.
- Check is the SFM for this workpiece material is within an acceptable range.

$$SFM = \frac{(\pi D) \times RPM}{12}$$

$$SFM = \frac{(\pi 2) \times 573}{12} = 300 ft / min$$

- 300 SFM would be too high for rough machining Ti 6Al4V.
- An acceptable range is 130 to 150 ft/min.
- Cutter would get too hot and prematurely fail.

- What recommendation would you make?

- Lock in an acceptable SFM of 140 ft/min
- Calculate the RPM.

$$RPM = \frac{SFM \times 12}{(\pi D)}$$

$$RPM = \frac{140 \times 12}{(\pi 2)} = 267 rev / min$$

- Recommend that this tool should be run at 267 RPM instead of 573 RPM.
- Request a change in program to the correct RPM.

Figure 80. Example 2 of SFM Calculations.

12.2 CHIP THICKNESS - CHIPLOAD - ADVANCE PER TOOTH (APT) & ADVANCE PER REVOLUTION (APR)

Before we talk about Feed, it is important to understand Chip Thickness. Very simply put, as shown in ***Figure 81***, it is the maximum thickness of a single chip. Chip Thickness, Chipload, and Advance Per Tooth (APT) are all industrial nomenclature that are used interchangeably.

As discussed earlier in Chapter 3, the APT is one of the dominant control factors for IMPACT in the cut, and consequently related to the toughness characteristics of the cutting tool. If the APT is higher than what the material of the cutting tool is designed to accommodate, it will result in higher than normal impact in the cut, and excessive chipping type wear of the cutting tool. This will eventually lead to premature or even catastrophic failure of the cutting tool.

The Advance Per Revolution (APR) is simply the distance travelled by the cutter over one complete revolution, as shown in ***Figure 81***. A cutting tool usually has multiple effective cutting edges (#T), each creating a single chip as they advance by their individual APT. Therefore:

$$APR = APT \times \#T \qquad [Units: in / rev]$$

In many ways, the APR is simply a visualizing tool to aid in the calculation of Feedrate in milling. However, special attention will be given to APR when it comes to drilling calculations in a later section.

Advance per Tooth (APT) is the distance travelled between two consecutive cutting edges at the center line of the cutter in the direction of advancement. It is also commonly referred to as the Chipload.

APT = Advance Per Tooth

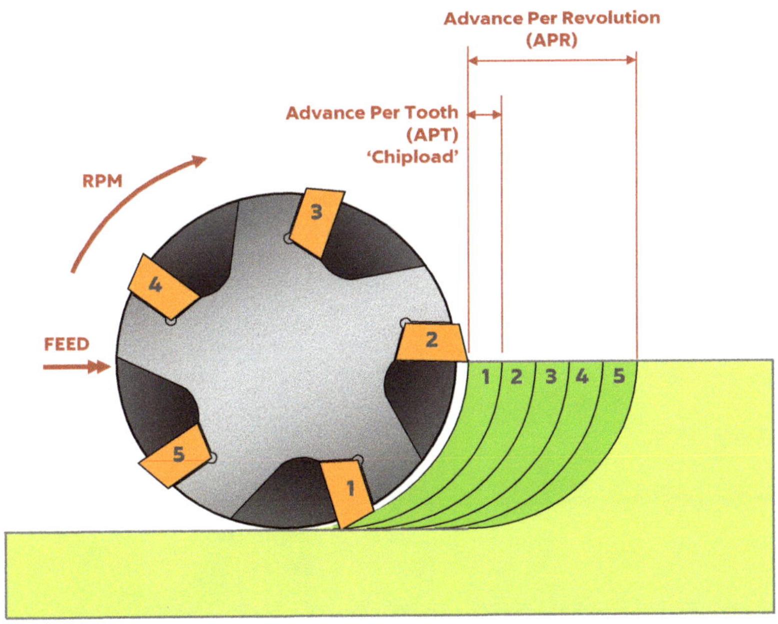

Advance per Revolution (APR) is the distance travelled by the cutting tool in one revolution.

APR = Advance Per Revolution

$$APR = APT \times (\#Teeth)$$

Figure 81. Advance per Tooth APT (Chipload), & Advance per Revolution APR.

12.3 FEEDRATE - INCHES PER MINUTE (IPM)

The Feedrate of a cutting tool is the resultant output of advancing the cutting tool through the workpiece. The best way to visualize the components of Feedrate is shown in ***Figure 82***, and by breaking it down by its various components.

- Every cutting edge advances by preplanned distance to make a single chip, and we have identified this as the Chipload (or Advance Per Tooth, APT).
- Next, based on the number of Effective Cutting Edges (or Number of Teeth, #T), each making their own individual chip, the cutter advances a known distance over one revolution (Advance Per Revolution, APR).

$$APR = APT \times \#T \qquad [Units: in]$$

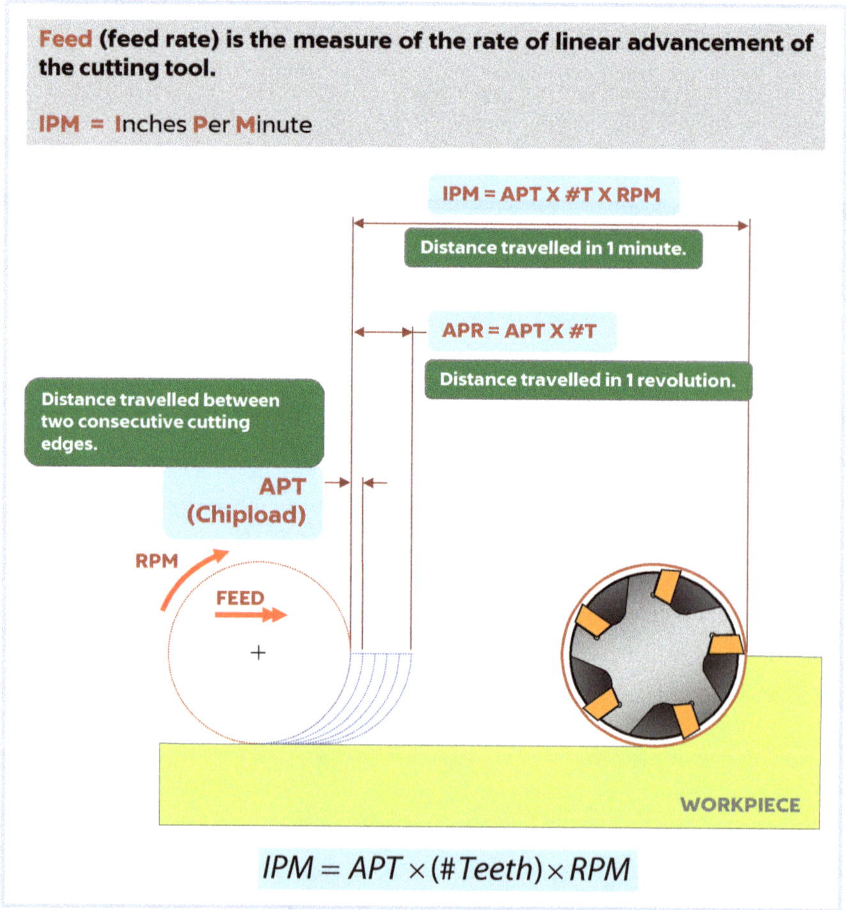

Figure 82. Feedrate - Inches Per Minute (IPM).

- Historically the Feedrate of a cutting tool is identified as how far the cutter would travel in one minute. In other word, how many revolutions a cutter completes in one minute (aka RPM).

$$IPM = APR \times RPM \qquad [Units:\ in/min]$$

$$IPM = APT \times \#T \times RPM \qquad [Units: in/min]$$

It is important to note that even though the APT is one of the main control factors for Impact in the cut, as well as the mathematical driver for IPM calculations, it is almost never seen numerically in an NC program or at a machine's control panel. It is IPM that show up in the NC code outputs, and often read on machine controllers as well.

12.3.1 LOCK THE APT & CALCULATE THE IPM

To close out the discussion on Feedrate, let's observe ***Figure 83***. Since Advance Per Tooth is the bite of a cutting edge, it has the most direct correlation to impact and edge pounding. Therefore, higher the APT, higher the impact in the cut. When focusing on APT, one needs to keep an eye on the chipping type wear of the cutting edge. Consequently, toughness is the material property of the cutting tool that resists chipping, and needs to be kept a watchful eye on.

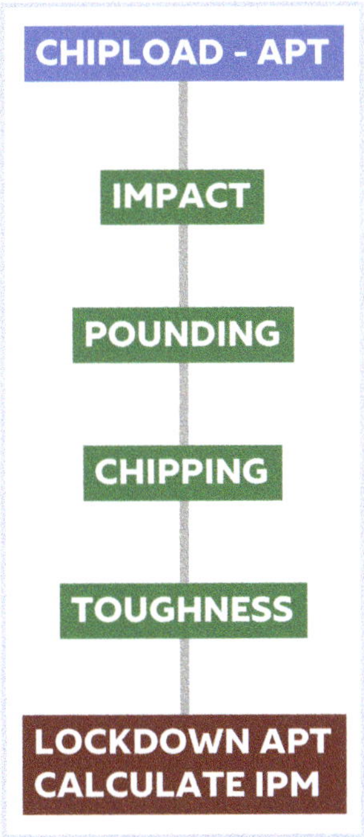

Figure 83. Chipload & its Variables.

A Best Practice: Chipload is predominantly responsible for the impact in the cut. For each workpiece material, we want to make sure that this impact signature remains just about the same throughout the spectrum of cutters that are responsible for machining the part. Depending on the size of the cutter, they often arrive with a manufacturer's recommended range of usable APT for a given workpiece material. Therefore, it is important to lock in the Advance Per Tooth (APT) for a given cutter and calculate the IPMs (***Figure 83***).

The example shown in ***Figure 84*** walks you through step-by-step scenario of calculating IPM for a given recommended APT in a known workpiece material.

- You are finish machining an exotic material, whose material properties of hardness and toughness require it to be machined at a very narrow chipload window of 0.002" to 0.003". It also requires the SFM to be at 110 ft/min. The tool to be used is a 1.25" DIA, 5 flute solid carbide endmill. What feedrate should this tool be programmed at?

- Recognize the known: DIA = 1.25", #T = 5, SFM = 110 ft/min, APT = 0.002" to 0.003".
- Need to calculate the Feedrate or IPM.

$$IPM = APT \times (\#Teeth) \times RPM$$

- However before we can calculate feedrate we need to know RPM.

$$RPM = \frac{SFM \times 12}{(\pi D)}$$

$$RPM = \frac{110 \times 12}{(\pi \times 1.25)} = 336 rev/min$$

- Now that we know the RPM, we can proceed to calculate the IPM.
- Lets decide to use an average APT of 0.0025"

$$IPM = APT \times (\#Teeth) \times RPM$$

$$IPM = 0.0025 \times 5 \times 336 = 4.2 in/min$$

- The tool should be programmed at 4.2 in/min.

Figure 84. Example of Feedrate (IPM) Calculation.

12.4 AN IMPORTANT RECOGNITION OF SPEED & FEED IN DRILLING

Although for the sake of this text we have used milling as the main method to explain the primary machining parameters, there are some unique differentiation factors that need to be addressed in drilling.

12.4.1 SFM & RPM IN DRILLING

A drill is a common tool used in hole making. The differentiation of a drill over a milling cutter, is that the drill cuts all the way from its center to it diametric periphery. To run a drill, the RPM or SFM calculation is identical to the one used in milling.

$$SFM = \frac{(\pi \times DIA).RPM}{12} \qquad \text{OR} \qquad RPM = \frac{SFM \times 12}{(\pi \times DIA)}$$

For these calculations, we use the max DIA of the drill. It should however be noted that at the very center point of a drill the SFM is zero. Essentially, as a drill is making a hole in the workpiece, its very central point is not really cutting in the traditional sense. Rather it is simply pushing into the workpiece material.

12.4.2 APR & IPM IN DRILLING

Notice that this section is titled "*APR & IPM*" in drilling, and not "*APT & IPM*". To best wrap our minds around this differentiation in drilling, it is important to understand the number of effective cutting edges on a drill as compared to that on a milling cutter. Almost all drills cutting edges culminate into a single point at the center, ***Figure 85***. As we spread out from this central tip, the drill splits into its various flutes. The main purpose of flutes is to guide the chips out of the hole being made. It would be wrong to count the number of flutes on a drill, same as the number of effective cutting edges. **At the center of any drill it is always one effective cutting edge**.

Depending on who you talk with, and the level of machining experience the person has, it could be erroneous to start talking Advance Per Tooth (APT). You never know if the person has considered the drill to be one effective cutting edge, or is mistakenly count the number of flutes. Therefore, it is always safe to ask or talk in terms of Advance Per Revolution (APR) when using a drill. This simple little personal rule has saved me many an embarrassing, catastrophic tool failures and potential scrapped parts. Therefore, in drilling the Feedrate (IPM) is calculated as:

$$IPM = APR \times RPM \qquad [Units:\ in/min]$$

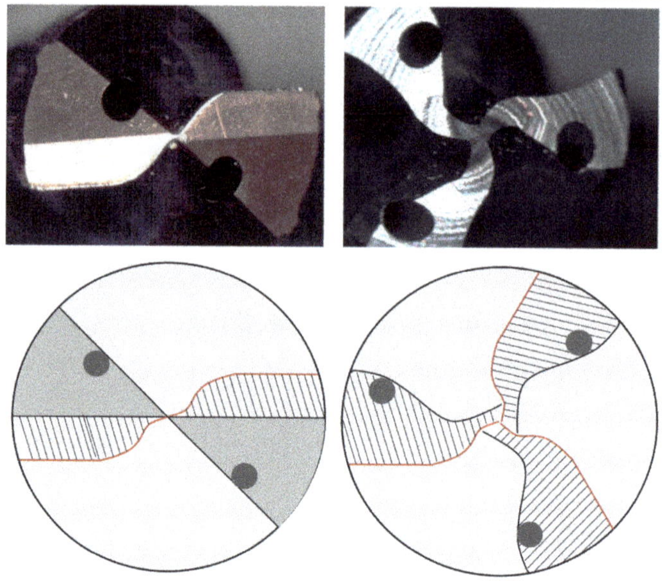

$$IPM = APT \times (1\ \text{Edge at Tip}) \times RPM$$

$$IPM = APR \times RPM$$

Figure 85. Calculating Feedrate for a Drill.

12.5 HIGH VELOCITY MACHINING

High Velocity Machining is also referred to as Shallow and Fast machining. There are cases where it is necessary to drop the Feedrate IPM in the machining process due to feature complexity. Some of these conditions could include thin walls, complex contoured geometry, machine rigidity, difficult to reach features, etc. Slowing down the traditional Speeds & Feedrates could yield a significant sacrifice in cycle times. This cycle time can be gained back by implementing High Velocity Machining techniques.

As the name implies, the user would shallow up the WOC, and consequently be able to run the RPM faster. The higher the RPM, the higher the IPM. A typical rule of thumb in High Velocity Machining is to engage the WOC at 10% (or less) of the diameter of the cutter. Consequently, the cutting edge makes a much smaller chip as compared to a traditional WOC engagement. This yields a much cooler heat signature on the overall life of the cutter. Therefore, in such a condition we are able to turn up the Surface Speed SFM (consequently RPM), and in turn we are able to yield a higher Feedrate IPM.

NOTE: High Velocity Machining does not necessarily yield higher Material Removal Rates MRR (discussed in next chapter). It is simply a technique that allows you to gain back some of the losses incurred from otherwise slowing down of a traditional velocity machining process that are necessitated by the part & feature limitations.

The actual implementation of High Velocity Machining is better understood after studying the concept of Chip Thinning Multiplier CTM in the following sections.

12.6 CHIP THINNING MULTIPLIER

The 'Chip Thinning Multiplier' is more so an industry slang nomenclature, rather than an accurate representation of its intended purpose. The chip is not truly thinning down. It is however a mathematical multiplier of the ratio of the Actual Diameter DIA to the Effective Diameter D_{Eff}.

Up to this point the calculations for Feedrate have been illustrated with the WOC engagement exactly at the center point of the cutter. (i.e. WOC = DIA/2), as seen in ***Figure 86*** (top). Consequently, this is also where the maximum chip thickness can occur. The basic Feedrate calculation of IPM=APT X #T X RPM, only works if the start of the chip only occurs at this WOC location. Recognize that the RPM in this equation is calculated as:

$$RPM = \frac{SFM \times 12}{\pi \times DIA}$$

Here the Diameter (DIA) is the same as the Effective Diameter (DIA_{Eff}).

As seen in ***Figure 86*** (bottom), the WOC is significantly less than half the diameter of the tool. Consequently, only a portion of the lower sector of the cutter is engaged in the cut. The chord formed at the location of WOC helps us visualize the Effective Diameter (D_{Eff}). Here the Effective Diameter is significantly less than the Actual Diameter of the tool. Now the basic Feedrate IPM calculation need to take into account the RPM that is calculated by D_{Eff}:

$$RPM = \frac{SFM \times 12}{\pi \times DIA_{Eff}}$$

The complete geometric & algebraic breakdown on calculating D_{Eff} is shown in ***Figure 87***, and works out to:

$$DIA_{Eff} = 2\sqrt{WOC.(DIA - WOC)}$$

A user certainly has the option to distinctly calculate the D_{Eff} and use it to calculate the RPM, which then gets plugged into the IPM calculation. In which case, there is no real need to calculate a 'Chip Thinning Multiplier'. However, the subsequent section will show the user-friendliness of simply calculating the 'Chip Thinning Multiplier' and applying it to the classically calculated IPM.

> **NOTE:** A very key point to recognize is that the ratio of maximum chip thickness WOC=DIA/2 to the chip thickness where WOC < DIA/2, is directly proportional to the ratio of Diameter to the Effective DIA. This ratio is essentially the Chip Thinning Multiplier (CTM).

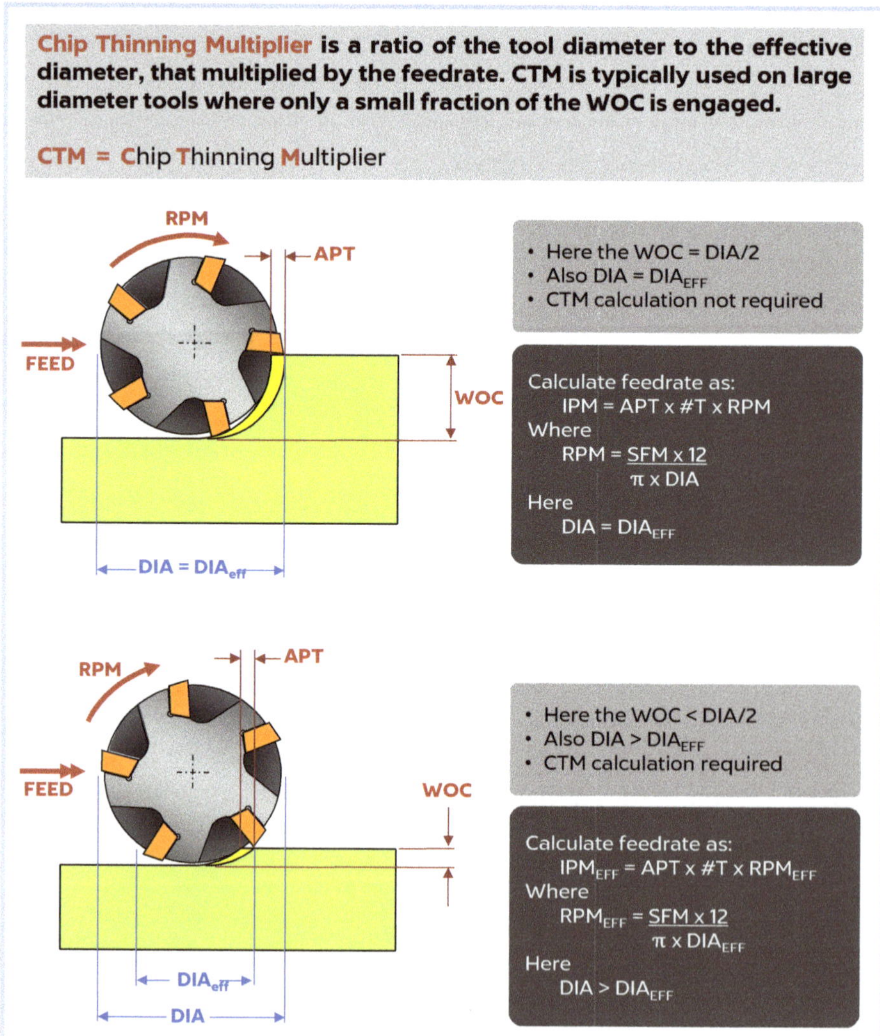

Figure 86. Chip Thinning Multiplier.

$$CTM = \frac{\text{Chip Thickness at } WOC = \frac{DIA}{2}}{\text{Chip Thickness at } WOC < \frac{DIA}{2}} = \frac{DIA}{DIA_{Eff}}$$

$$CTM = \frac{DIA}{2\sqrt{WOC.(DIA - WOC)}}$$

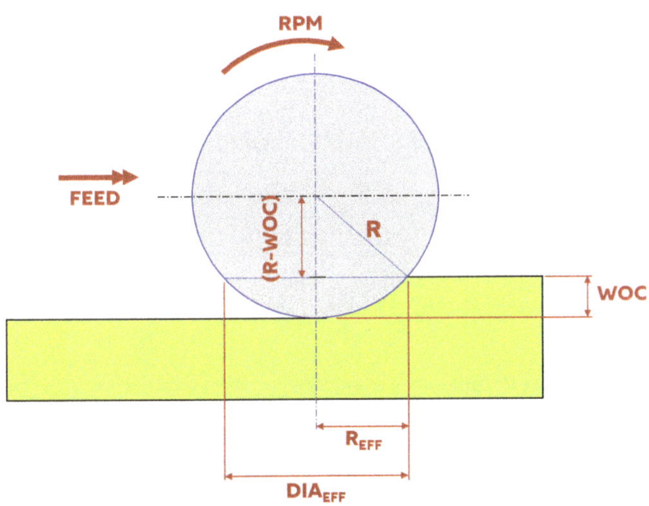

DERIVE DIA_{EFF}

$R^2 = R_{EFF}^2 + (R - WOC)^2$

$R_{EFF}^2 = R^2 - (R - WOC)^2$

$R_{EFF} = [R^2 - (R^2 - 2R.WOC + WOC^2)]^{1/2}$

$R_{EFF} = [R^2 - R^2 + 2R.WOC - WOC^2]^{1/2}$

$R_{EFF} = [2R.WOC - WOC^2]^{1/2}$

$R_{EFF} = [DIA.WOC - WOC^2]^{1/2}$

$R_{EFF} = [WOC(DIA - WOC)]^{1/2}$

$DIA_{EFF} = 2[WOC(DIA - WOC)]^{1/2}$

$CTM = DIA / DIA_{EFF}$

$$CTM = \frac{DIA}{2[WOC(DIA - WOC)]^{1/2}}$$

Calculate feedrate as:

IPM = APT x #T x RPM

Where

$$RPM = \frac{SFM \times 12}{\pi \times DIA}$$

$$IPM_{EFF} = IPM \times CTM$$

Figure 87. Chip Thinning Multiplier.

12.6.1 A USER-FRIENDLY APPROACH TO IMPLEMENTING CTM

The in-depth derivation of Chip Thinning Multiplier is not really necessary for implementing it. This section shows a step-by-step approach to implementing CTM into your IPM calculation:

STEP 1: Calculate the IPM by using the basic classic equation:

$$IPM = APT \times \#T \times RPM$$

- Here use Actual Diameter of cutter to calculate RPM.
- Use desired Chipload (APT) at center line of cutter (i.e. at WOC=DIA/2).

STEP 2: Calculate the Chip Thinning Multiplier using the equation:

$$CTM = \frac{DIA}{2\sqrt{WOC.(DIA - WOC)}}$$

- Recognize that the CTM will always be a value greater than 1.

STEP 3: Calculate the final Feedrate (IMP_{Final}) by multiplying the IPM calculated in Step 1 by the CTM from Step 2:

$$IPM_{Final} = IPM \times CTM$$

NOTE: In closing, CTM is simply a multiplier to the IPM. It ensures that when you are a running light WOC (<DIA/2), that the desired Chipload is maintained at the point of chip formation. It does so by appropriately cranking up (multiplying) the IPM at the cutter centerline (DIA/2).

CAUTION: If you just go by the mathematics of it, as soon as WOC is less than DIA/2, you will always have a CTM value that is greater than 1. Does that mean that you should incorporate it into your IPM calculation every time? Personally, if the CTM is between 1 and 1.5, there is not much benefit to implementing it. If the CTM is between 1.5 and 2, it may be worthy of considering for implementation (especially in soft metal like applications, like Aluminum). If the CTM is greater than 2, it should definitely be considered for implementation. These judgement calls become a lot more solidified after a firm understanding of Material Removal Rates (MRR) in the next chapter.

12.6.1.1 CTM IMPLEMENTATION WITH NO MATH

If for some reason the user chooses to simply stay away from the math of it all, but still wants the productivity benefits of an increased Feedrate (IPM), this section talks about a real life & practical approach. Simply get into the cut at a machine with an IPM that is calculated at a nominal Chipload (APT). If the WOC is very shallow, it is quite likely that the chips are coming out extremely thin, and in some cases are even undesirable. Next, at the machine's controller simply turn up the IPM in increments of 5% or 10%. The process will give you feedback. Listen to the sound of the cut, and watch the shape & color of the chips. Keep increasing the IPM based on your comfort level of cutter sound, and shape of chips. This is where the art of metal cutting, & experience will come into effect. If you are a beginner, ask someone with experience to lend you feedback on the sound of the operation and the desired shape of the chip. This cranking up of the Feedrate (IPM) has been done by many generations of machinists, and can prove to be significantly reliable & is often a well-rounded solution for increasing efficiency. The math of is this practical art is the Chip Thinning Multiplier calculation.

CAUTION: In the history of machining, there have been numerous disastrous cases where Chip Thinning Multipliers (CTM) have been implemented in Feedrate calculations simply based on their math merit. Feedback from the operator community is crucial, and should be a heavily weighted criterion. This is especially true in cases such as hard metals (Titanium, steels) machining with High Velocity techniques (shallow WOC & fast RPM), where the CTM can calculate of to 3X, 4X, or even 5X. Their input on the sight, sound, & performance of a process can easily counteract the implementation of CTM (or the magnitude of it). Some of these extraneous factors may include: operation sound, vibrations, chip color, heat in cut, part warpage issues, chip packing, tool wear, etc.

The proper implementation of Chip Thinning Multiplier (CTM) can yield a significant cycle time benefit. However, its implementation should be practical, and based on well-rounded production input. Emphasis should be given to 'repeatable reliable performance'.

12.7 TURNING SPEEDS & FEEDS

Now that the derivation of Speed & Feed is adequately explained for milling, the equivalent turning can easily be visualized. These are clearly derived in ***Figure***

88. It should be recognized that in turning there is usually only one cutting edge. Therefore, the concept of Chipload (APT) as seen in milling for Feedrate (IPM) calculations, is not as relevant in turning. Rather, the more relevant term for turning Feedrate calculation is Advance Per Revolution (APR).

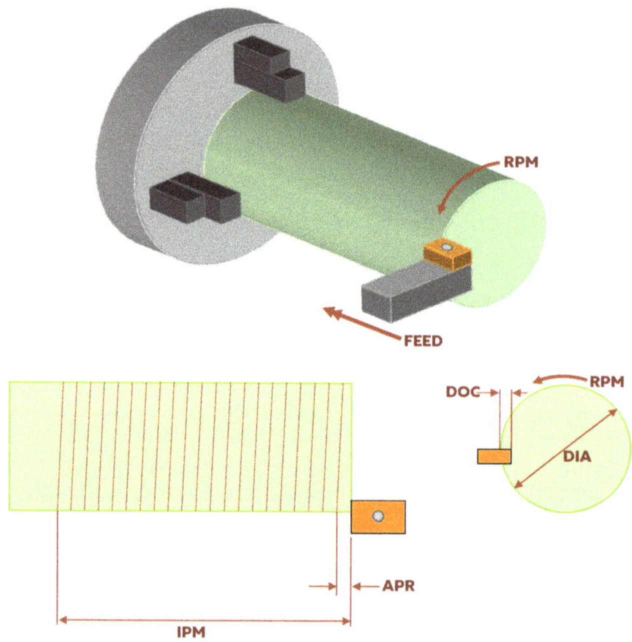

$$SFM = \frac{\pi \times DIA \times RPM}{12} \qquad IPM = APR \times RPM$$

Figure 88. Turning Speeds & Feeds.

13

MATERIAL REMOVAL RATE

The concept of Material Removal Rate (MRR) is arguably one of the truest measures of efficiency in a machining process. Thus far all the concepts covered in this text are leading up to empower the user to correctly understand Material Removal Rate. The machining world is saturated with perceptions of machining efficiencies, and 'wow-factors' that may or may not pay off in real life production. In this world of machining, a person's opinion can get easily swayed by impressive concepts such as: High Speed, Large Cutter sizes, High Density Cutters, Power, Torque, multi-axis machines, high pressure coolant, multi-spindles, robotics, cryogenics, and countless other terminologies with their inherent 'wow-factors'. Here we will find that an MRR calculation is one of the primary litmus tests that would allow user to make a data-driven decision on the true efficiency of a process. Of course, the highest value of MRR is not always the best solution. It has to stand balanced against tool life, cost, safety, and repeatable reliable performance.

13.1 CALCULATING MATERIAL REMOVAL RATE (MRR) FOR A CUTTING TOOL

Simply put, MRR represents 'How quickly can we make chips?'. It is actually the volume of chips created over one minute. The MRR units of measurement are in^3/min. As shown in ***Figure 89*** and ***Figure 90*** the MRR calculation can be visualized as follows:

- The DOC and WOC represents the area of a rectangular plane created by the engagement of the cutting tool in the workpiece (***Figure 89***).
- Next, we ask ourselves, how far would this plane travel over one minute? This distance is represented by the Feedrate or Inches Per Minute (IPM). (***Figure 90***).
- Therefore, the MRR in milling is calculated by a simple equation:

$$MRR = DOC \times WOC \times IPM \qquad [\text{Units: } in^3 / min]$$

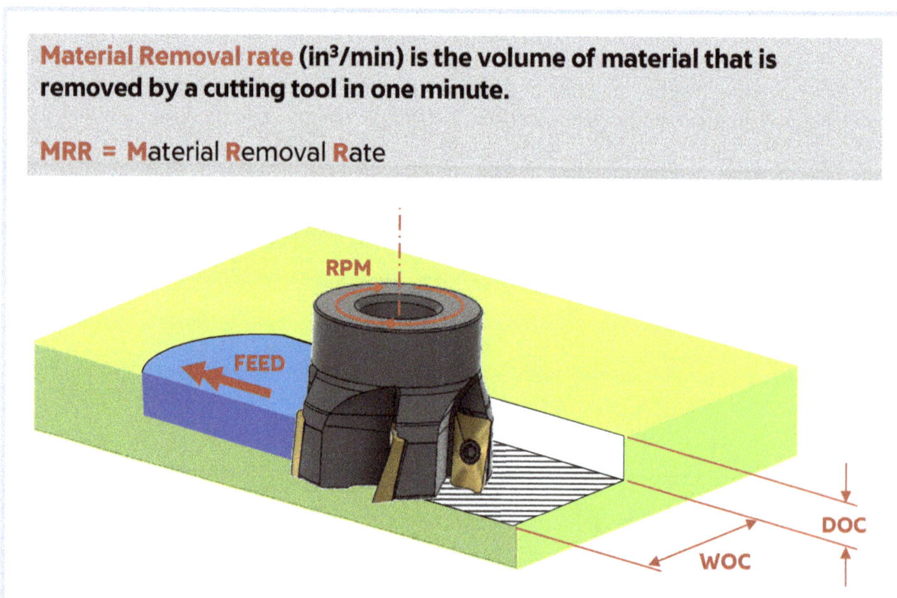

Figure 89. Material Removal Rates.

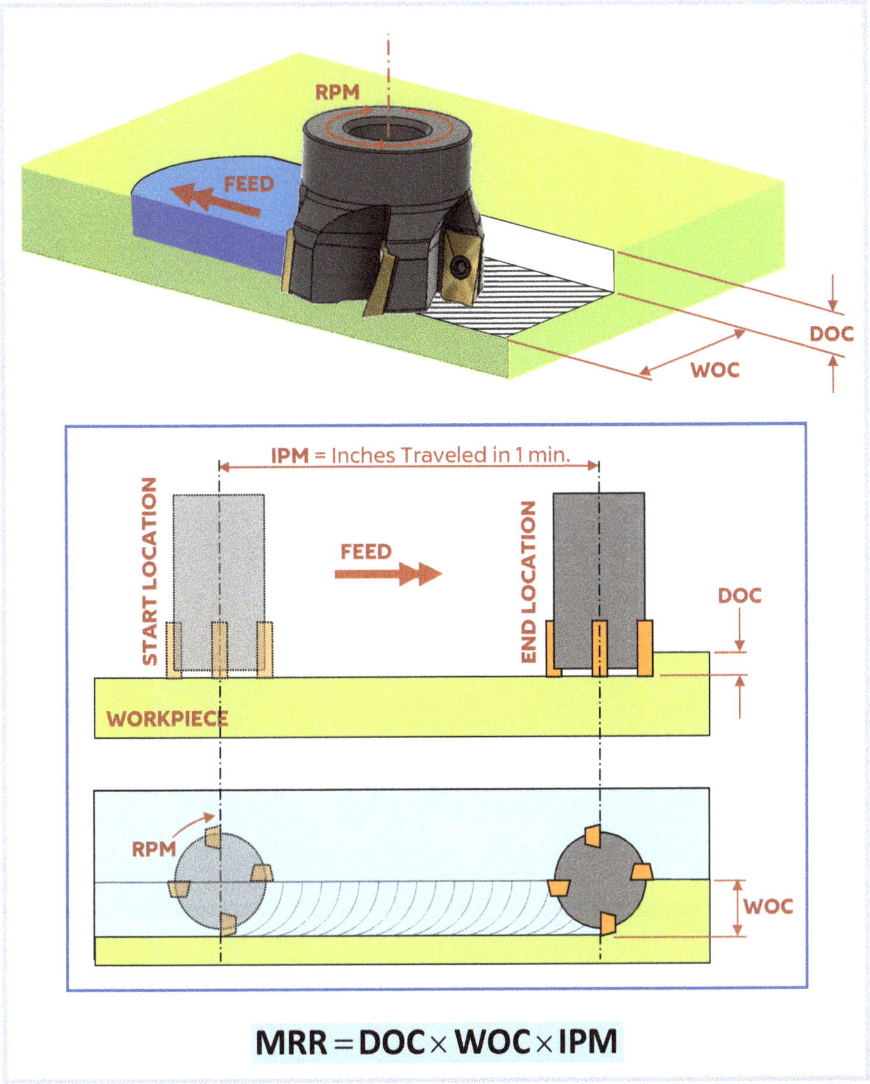

Figure 90. Calculating Material Removal Rate in Milling.

13.2 SAMPLE CALCULATIONS OF MRR

The following subsections walk the user through some real-life calculations of MRR for Solid Carbide Endmills, as well as Indexable roughing cutters. It showcases how small adjustments in Speeds & Feeds and significantly affect MRRs.

13.2.1 MRR FOR A SOLID CARBIDE ENDMILL IN TITANIUM

EXAMPLE 1.
WORKPIECE: Titanium
TOOL: Solid Carbide Endmill
DIA = 1"
#T = 5 flutes
SFM = 140 ft/min
APT = 0.003"
DOC = 2"
WOC = 0.050"
CALCULATE: RPM, IPM, & MRR

RPM = 535 rev/min
IPM = 8 in/min
MRR = 0.8 in^3/min

EXAMPLE 2.
WORKPIECE: Titanium
TOOL: Solid Carbide Endmill
DIA = 1"
#T = 5 flutes
SFM = 200 ft/min
APT = 0.003"
DOC = 2"
WOC = 0.050"
CALCULATE: RPM, IPM, & MRR

RPM = 764 rev/min
IPM = 11.5 in/min
MRR = 1.1 in^3/min

EXAMPLE 3.
WORKPIECE: Titanium
TOOL: Solid Carbide Endmill
DIA = 1"
#T = 5 flutes
SFM = 200 ft/min
APT = 0.005"
DOC = 2"
WOC = 0.050"
CALCULATE: RPM, IPM, & MRR

RPM = 764 rev/min
IPM = 19.1 in/min
MRR = 1.4 in^3/min

Figure 91. Sample Calculation – Solid Carbide Endmill.

EXAMPLE 1.
WORKPIECE: Titanium
TOOL: Indexable Endmill
DIA = 3"
#T = 5 flutes
SFM = 130 ft/min
APT = 0.004"
DOC = 4"
WOC = 1.25"
CALCULATE: RPM, IPM, & MRR

RPM = 166 rev/min
IPM = 3.3 in/min
MRR = 16.6 in³/min

EXAMPLE 2.
WORKPIECE: Titanium
TOOL: Indexable Endmill
DIA = 3"
#T = 5 flutes
SFM = 150 ft/min
APT = 0.004"
DOC = 4"
WOC = 1.25"
CALCULATE: RPM, IPM, & MRR

RPM = 191 rev/min
IPM = 3.8 in/min
MRR = 19.1 in³/min

EXAMPLE 3.
WORKPIECE: Titanium
TOOL: Indexable Endmill
DIA = 3"
#T = 5 flutes
SFM = 150 ft/min
APT = 0.006"
DOC = 4"
WOC = 1.25"
CALCULATE: RPM, IPM, & MRR

RPM = 191 rev/min
IPM = 5.8 in/min
MRR = 28.6 in³/min

Figure 92. Sample Calculation – Indexable Cutter.

EXAMPLE 1.
WORKPIECE: Titanium
TOOL: Indexable Endmill

DIA	= 3"
#T	= 5 flutes
SFM	= 140 ft/min
APT	= 0.006"
DOC	= 4"
WOC	= 1.25"

CALCULATE: RPM, IPM, & MRR

RPM = 178 rev/min
IPM = 5.3 in/min
MRR = 26.7 in^3/min

EXAMPLE 2.
WORKPIECE: Titanium
TOOL: Button Cutter

DIA	= 4.5"
#T	= 6 flutes
SFM	= 500 ft/min
APT	= 0.002"
DOC	= 0.25"
WOC	= 3.75"

CALCULATE: RPM, IPM, & MRR

RPM = 424 rev/min
IPM = 6.8 in/min
MRR = 4.8 in^3/min

EXAMPLE 3.
WORKPIECE: Aluminum
TOOL: Aluminum Router

DIA	= 2.0"
#T	= 3 flutes
SFM	= 13090 ft/min
APT	= 0.006"
DOC	= 0.75"
WOC	= 0.6"

CALCULATE: RPM, IPM, & MRR

RPM = 25,000 rev/min
IPM = 450 in/min
MRR = 202 in^3/min

Figure 93. Sample Calculation – Indexable Endmill vs Button Cutter.

13.3 TURNING - MATERIAL REMOVAL RATE

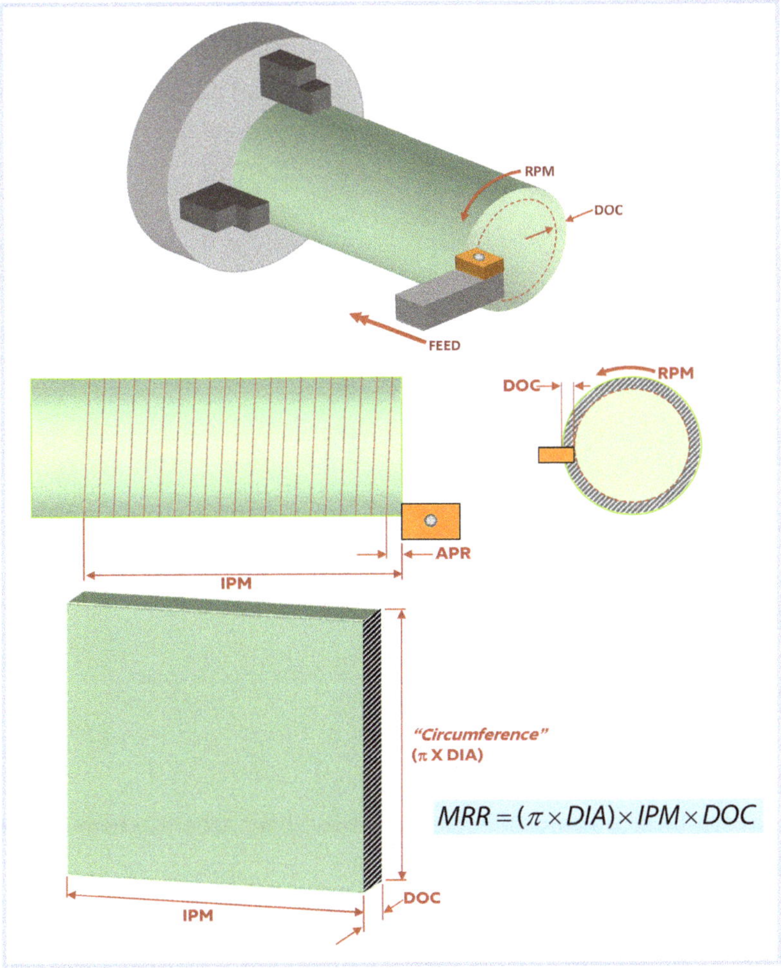

Figure 94. Turning Material Removal Rate Calculation.

13.4 DRILLING – MATERIAL REMOVAL RATE

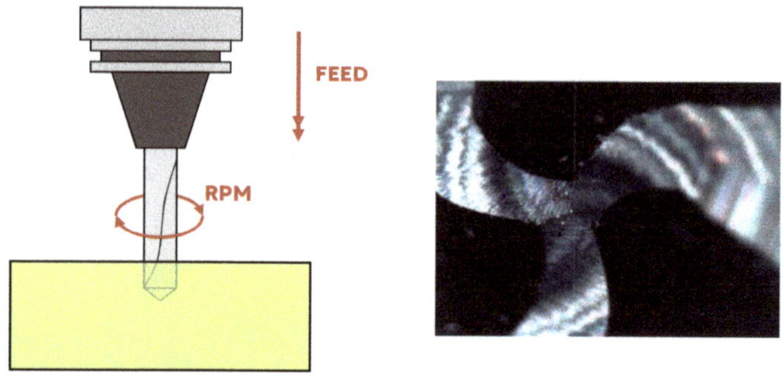

CAUTIONS:

- In a drill the center point is where all the flutes connect.
- At the center point, since the **DIA** is virtually zero, the **SFM** is zero. Therefore the center of the drill simply pushes through the workpiece material.
- When doing **FEEDRATE (IPM)** calculations, it is always a safe approach to consider all drills as one effective cutting edge. Therefore do not consider **CHIPLOAD (APT)** in a drill. It is always a safe approach to use Advance Per Revolution **(APR).**

$$SFM = \frac{\pi \times DIA \times RPM}{12} \qquad IPM = APR \times RPM$$

MATERIAL REMOVAL RATE is the area of the hole multiplied by the **FEEDRATE (IPM).**

$$MRR = \left(\pi \times \left(\frac{DIA}{2} \right)^2 \right) \times IPM$$

Figure 95. Drilling Material Removal Rate Calculation.

13.5 CALCULATING PROCESS MATERIAL REMOVAL RATE ($MRR_{PROCESS}$)

Certainly, calculating MRR for a single cutting tool by multiply the DOC, WOC, and IPM is accurate mathematically. If you are in the business of manufacturing cutting tools and want to compare your latest new product to an earlier generation one, this calculation is absolutely the correct way to do so. However, if you are in a production world, while machining a part you may not have the luxury of a constant engagement or Feedrate. Most real-life parts require the DOC and WOC to be varied. Additionally, the IPMs need to be adjusted based on the complexity of the feature being machined (e.g. corners, interruptions, entry, and exit). It should also be recognized, that to machine a complete part it takes a variety of cutters with varying diameters, number of edges, and flute lengths. In such a case, it would be extremely erroneous to showcase the max MRR of a single cutter doing work over a finite amount of time. For such an analysis where the total process has to be taken into account, there is a slightly alternate way to look at MRR. We will refer to this as the Process MRR.

At its very core the Process Material Removal Rate can be viewed as:

$$MRR_{PROCESS} = \frac{Volume\ of\ Chips}{TIME}$$

Let's now breakdown the two components of this equation (Volume, & Time).

13.5.1 VOLUME OF CHIPS

There is a simple way to calculate the volume of chips without knowing anything about the actual machining process or cutter engagement. This simple technique could prove very useful to industrial engineers, management, and any other team member who may or may not be well versed in the sciences of machining.

- In almost all production environments we know the weight of the forging (or weight of starting material block), and the weight of the finished part. The difference between the two is consequently the weight of the chip. These weights could be determined by simply:
    - Weighing a forging, and weighing a finished part.
    - Using the computer modeling software to forecast the weights.

$$Weight\ of\ Chips = Weight\ of\ Forging - Weight\ of\ Finished\ Part$$

- Next, we know the material of the workpiece being machined. Hence, we know the density of the material. From this we can calculate the Volume of Chips as:

$$Volume\ of\ Chips = \frac{Weight\ of\ Chips}{Density\ of\ Workpiece}$$

13.5.2 TIME

When it comes to the Time component of a machining process you have to ask yourself, what amount of the total processes time is important to you. Why and why not? Some relevant examples of measures of process time are:

- The time a single cutting tool is in the cut.
  (**For Example:** If you were a cutter manufacturer trying to evaluate the exact tool life of your new product.)
- The cutting and non-cutting time of the single cutter.
  (**For Example:** If you were an NC programmer trying to evaluate the efficiency of a unique cutter path strategy.)
- The time taken by all the cutters to make the finished part.
  (**For Example:** If you were tasked by your management to reduce the total cycle time of a machining process to create capacity on that machine.)
- The time the workpiece is shuttled into a machine, till the time it comes out.
  (**For Example:** If you were a shop manager trying to increase the overall throughput of your particular process, or compare it to another equivalent one.)
- The time from when the workpiece arrives at your site, till the time the finished part is shipped out.
  (**For Example:** If you were the company owner or factory manager who is interested in the bottom line. After all, you only get paid when the product leaves you factory.)

All the above measures of time serve a purpose and are relevant to the task you are trying to accomplish, or are relevant questions you are trying to answer.

In summary, now that you have the Volume of Chips, and the measure of time that is relevant to your analysis, you can simply calculate the Process MRR by the earlier stated equation:

$$MRR_{PROCESS} = \frac{Volume\ of\ Chips}{TIME}$$

Material Removal Rate is an equalizer of real and perceived performance gains. The process MRR is true representation of machining, and can be used to compare similar machining operations of the same material. It is important not to forget about costs (i.e. Process MRR at what cost?). Finally, process MRR is aligned with synergy of operations, and can be a simple analysis that helps compare the efficiencies versus inefficiencies between similar and or dissimilar operations.

14

POWER, CUTTING FORCES, TORQUE, BENDING MOMENTS, & AXIS DRIVE RATINGS

This chapter focuses on some of the key drivers and parameters in the selection of an appropriate machine to accomplish the cutting task at hand. You could have a state-of-the-art cutter purchased, but these are rendered useless if you don't have enough machine behind them capable of pushing the cutting edges through the cut.

The manner in which Power, Torque, Cutting Force, & Bending Moments are calculated in metal cutting is slightly varied from the typical calculations used in physics. Although the principles are the same, the approach is varied. It is not the intent of this text to deep dive the physics-based understanding, but rather to provide a high-level summary of it. Stemming from Material Removal Rate (MRR), this text does however provide the continuity of calculating Power, Cutting Forces, Torque, and Bending Moments.

A slightly more in-depth understanding of the physics is shown in ***Figure 96***. Once again, the intent here is just provide a high-level summary of physics-based definitions, and possible to appease a curious mind.

MASS *Is a measure of how much matter an object contains. It is the total number of subatomic particles (electrons, protons, & neutrons) in the object.*

WEIGHT *Is mass multiplied by the pull of gravity.*

GRAVITY: *9.8 m/s², or 32 ft/s².*

FORCE = mass X acceleration.
Force applied to an object causes acceleration. An object's weight is a type of force that the earth exerts upon it.

Newton's Second Law: *Acceleration (a) of an object is directly proportional to the force (F), and inversely proportional to the object's mass (m).*

TORQUE = Force X Distance from center.
Is a force that tends to rotate an object. You generate a torque any time you apply force using a wrench.

NOTE: The object need not move when Torque is applied.

WORK = Force X Distance in the direction of the force.
Is the application of a force over a distance. However, the distance only counts if it is in the direction of the applied force.

Lifting a weight from the ground and putting it on a shelf is a good example of work. The force is equal to the weight of the object, and the distance is equal to the height of the shelf. If the weight were in another room, and you had to pick it up and walk across the room before you put it on the shelf, you didn't do any more work than if the weight were sitting on the ground directly beneath the shelf. It may have felt like you did more work, but while you were walking with the weight you moved horizontally, while the force from the weight was vertical.

Your car also does work. When it is moving, it has to apply a force to counter the forces of friction and aerodynamic drag. If it drives up a hill, it does the same kind of work that you do when lifting a weight. When it drives back down the hill, however, it gets back the work it did. The hill helps the car drive down.

Work is energy that has been used. When you do work, you use energy. But sometimes the energy you use can be recovered. When the car drives up the hill, the work it does to get to the top helps it get back down. Work and energy are closely related. The units of work are the same as the units of energy.

POWER *Is a measure of how quickly work can be done.*
By using a lever, you may be able to generate 200 ft-lb of torque. But could you spin that lever 3000 times per minute.
You can multiply the amount of torque by the rotational speed in order to find the power.
Another way to look at power is as a unit of speed combined with a unit of force.

ENERGY If power is like the strength of a weightlifter, energy is like his endurance. *Energy is a measure of how long we can sustain the output of power, or how much work we can do.* Power is the rate at which we do the work. One common unit of energy is the kilowatt-hour (kWh). You learned in the last section that a kW is a unit of power. If we are using one kW of power, a kWh of energy will last one hour. If we use 10 kW of power, we will use up the kWh in just six minutes. There are two kinds of energy: **potential** and **kinetic**.

POTENTIAL ENERGY is waiting to be converted into power. Gasoline in a fuel tank, food in your stomach, a compressed spring, and a weight hanging from a tree are all examples of potential energy.

The human body is a type of energy-conversion device. It converts food into power, which can be used to do work. A car engine converts gasoline into power, which can also be used to do work. A pendulum clock is a device that uses the energy stored in hanging weights to do work.

When you lift an object higher, it gains potential energy. The higher you lift it, and the heavier it is, the more energy it gains. For example, if you lift a bowling ball 1 inch, and drop it on the roof of your car, it won't do much damage (please, don't try this). But if you lift the ball 100 feet and drop it on your car, it will put a huge dent in the roof. The same ball dropped from a greater height has much more energy. So, by increasing the height of an object, you increase its potential energy.

Let's go back to our experiment in which we ran up the stairs and found out how much power we used. There is another way to look at how we calculated our power: We calculated how much potential energy our body gained when we raised it up to a certain height. This amount of energy was the work we did by running up the stairs (force * distance, or our weight * the height of the stairs). We then calculated how long it took to do this work, and that's how we found out the power. Remember that power is the rate at which we do work.

The formula to calculate the potential energy (PE) you gain when you increase your height is:

PE = Force * Distance
In this case, the force is equal to your weight, which is your mass (m) * the acceleration of gravity (g), and the distance is equal to your height (h) change. So the formula can be written:

PE = mgh

KINETIC ENERGY is energy of motion. Objects that are moving, such as a roller coaster, have kinetic energy (KE). If a car crashes into a wall at 5 mph, it shouldn't do much damage to the car. But if it hits the wall at 40 mph, the car will most likely be totaled.
Kinetic energy is similar to potential energy. The more the object weighs, and the faster it is moving, the more kinetic energy it has. The formula for KE is:
KE = 1/2*m*v2,
where m is the mass and v is the velocity.
One of the interesting things about kinetic energy is that it increases with the velocity squared. This means that if a car is going twice as fast, it has four times the energy. You may have noticed that your car accelerates much faster from 0 mph to 20 mph than it does from 40 mph to 60 mph. Let's compare how much kinetic energy is required at each of these speeds. At first glance, you might say that in each case, the car is increasing its speed by 20 mph, and so the energy required for each increase must be the same. But this is not so.

We can calculate the kinetic energy required to go from 0 mph to 20 mph by calculating the KE at 20 mph and then subtracting the KE at 0 mph from that number. In this case, it would be 1/2*m*202 - 1/2*m*02. Because the second part of the equation is 0, the KE = 1/2*m*202, or 200 m. For the car going from 40 mph to 60 mph, the KE = 1/2*m*602 - 1/2*m*402; so KE = 1,800 m - 800 m, or 1000 m. Comparing the two results, we can see that it takes a KE of 1,000 m to go from 40 mph to 60 mph, whereas it only takes 200 m to go from 0 mph to 20 mph.

There are a lot of other factors involved in determining a car's acceleration, such as aerodynamic drag, which also increases with the velocity squared. Gear ratios determine how much of the engine's power is available at a particular speed, and traction is sometimes a limiting factor. So it's a lot more complicated than just doing a kinetic energy calculation, but that calculation does help to explain the difference in acceleration times.

Figure 96. A Quick Physics Lesson

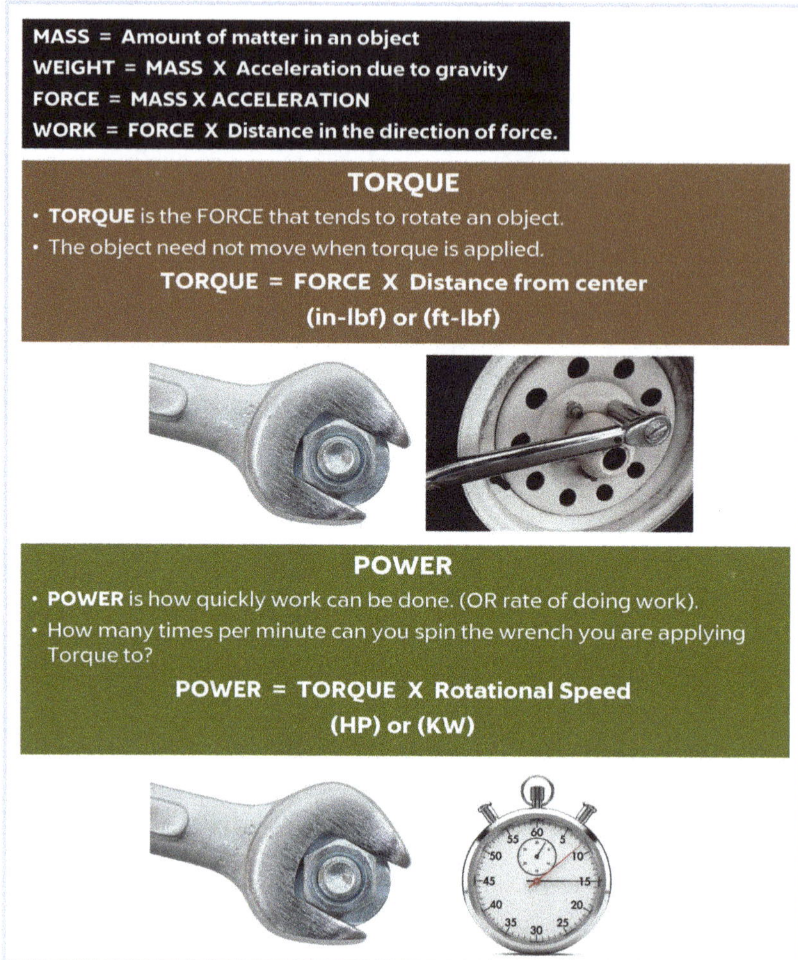

Figure 97. Summary Review of Power & Torque.

14.1 CALCULATING POWER IN METAL CUTTING

14.1.1 THE PHYSICS OF POWER

In physics Work is defined as the Energy transferred by a Force. Where:

$$Work = Force \times Distance$$

Power is defined as the rate of doing work

$$Power = \frac{Work}{Time}$$

Also,

$$Power = Force \times \ Velocity$$

14.1.2 CALCULATING POWER IN METAL CUTTING

In the world of machining Power is calculated as:

$$Power = MRR \times \ K \qquad [\text{Units}: \ Hp]$$

As explained in the earlier sections MRR brings the Volume of material to be removed, along with the Time component. K is the material factor (***Figure 98***).

14.1.2.1 MATERIAL FACTORS

The implementation of the Material Factor (K), is one of the primary differentiations in Power calculation, from the typical physics-based approach. Although you may have realized the volume of material to be removed, you still have to factor in the resistance that various material types will present during cutting. For example, a soft piece of plastic would provide very little resistance as compared to a hardened steel. Typically, the user would select K from a table (see ***Figure 99***). However, this text also shows a how these tables are created by industry in a lab environment. Additionally, if you do not have access to a full-scale fancy lab, don't worry, there are some simple techniques described that will allow you to build you own Material Factor tables from existing machines. All of this is best talked about after we discuss Cutting Forces in a later section.

POWER is the rate of doing work. It is usually measured in Horsepower (HP) or Kilowatts (KW). In the world of metal cutting we are mainly concerned with the power it takes the spindle to push though a cut in a given workpiece material.

One Horsepower is defined as *one horse lifting 550 pound by a distance of one foot for each second elapsed.* (NOTE: 1KW = 1.34HP)

$$HP = MRR \times MF$$

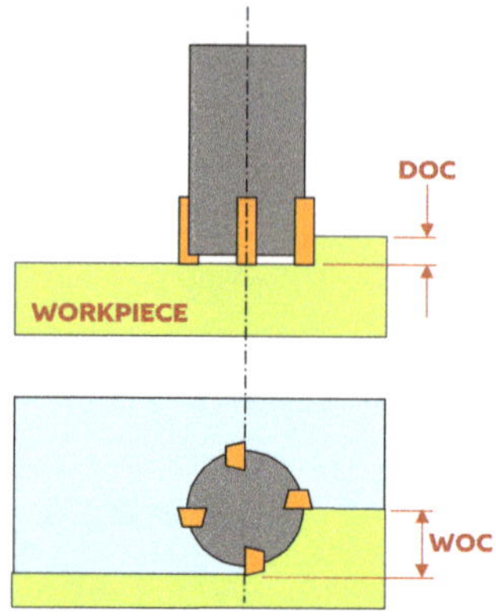

WHERE:
MF is the Material Factor in units HP min/in³.

NOTE: It is important to pay attention to the units when looking up MF in tables.

- If MF is HP min/in³ then multiply MRR by MF.
- If MF is in³/HP min, then divide MRR by MF.

Figure 98. Calculating Power in Metal Cutting.

14.1.2.1.1 Caution On Material Factors (K) Vs Machinability Factors (M) Vs Power Factors (P)

This cautionary note is nothing more than an awareness of nomenclature variations that commonly occurs in the metal cutting industry. Depending on who you talk to, folks in the trade use terms such as Material Factors, Machinability Factors, and Power Factors interchangeably. The cautionary note is that when you read off some of these tables the units could be reciprocals of each other, (HP min/in^3) **OR** (in^3/ HP min).

CAUTION: In this text, we multiply MRR (in^3/min) by K to calculate the Power (HP). Therefore, the units of K must be (HP min/in^3). If you should choose to obtain these values from an alternate handbook table, pay special attention to the units. If the units are represented as (in^3/HP min), then you must divide this MRR by this value to obtain the power. Bottom line, irrespective of what it is called (K, M, or P), pay attention to the units and then decide if you need to multiply or divide this value to the MRR.

MATERIAL	BRINELL HARDNESS	DESCRIPTION	INDUSTRY	MATERIAL FACTOR K (HP min/in^3)
Unalloyed Steels	110	Low Carbon (< 0.3%)	Constructional, structural, deep drawn, stamped products, pressure vessel steel & a variety of cast steels. General Uses: axles, shafts, tubes, forgings & welded constructions (1010, 1020 steel)	0.75
	150	Medium Carbon (< 0.6%)		0.8
	300	High Carbon (> 0.6%)		1.1
Low Alloy Steels	125 - 220	Not Hardened	General Uses: axles, shafts, structural steels, tubes and forgings. Automotive: con rods, cam shafts, cv-joints, wheel hubs and steering pinions (4340, 4140 steel).	0.93
	220 - 420	Hardened		1.22
High Alloy Steel	150 - 300	Annealed	General Uses: machine tool parts, dies, hydraulic components, cylinders and cutting tools (HSS)	1.1
	275 - 350	Hardened Tool Steel		1.7
Super Hard Steel	> 450	Hardened & Tempered	Components: transmission shafts, gear box housings, steering pinions, stamping dies.	1.7
Cast Iron (Malleable)	100 - 150	-	Components: axle bearings, track wheels, pipe fittings & high-strength gears.	0.45
	200 - 230	-		0.5
Cast Iron (Grey)	180	Low Tensile	Components: frying pans, engine blocks, cylinders for compressors, gears & gear box cases	0.48
	250	High Tensile, Alloyed		0.55
Cast Iron (Nodular)	160	Ferritic	Components: Hubs, tubing, rollers, exhaust manifolds, crankshafts, differential housings, bearing caps, bedplates, turbo charger housings, clutch plates & fly wheels.	0.45
	250	Pearlitic		0.77
Stainless Steel	150 - 270	Ferritic, Martensitic 13 - 25% Cr	Ferritic: shafts for pumps, turbines, steam and water turbines, nuts, bolts, hot water heaters, pulp & food processing industries. Martensitic: edges in cutlery steel, razor blades and surgical instruments, etc.	1.1
	150 -275	Austenitic > 8% Ni, 18 - 25% Cr	Chemical, pulp and food processing industries, & exhaust manifolds for airplanes	0.92
	275 - 425	Quenched & tempered, Martinsitic > 0.12% C	Chemical, food, construction, medical, cellulose & paper-making industries & in processes that include acids, & off-shore oil and gas industry	1.05
	150 - 450	Precipitation hardened steels		1.3
Heat Resistant Super Alloys [Fe, Ni, & Co based]	140 - 300	Annealed or solution treated	Aerospace engine & power gas turbines, oil & gas marine applications, medical joint implants, high corrosion resistant applications	1.4
	250 - 475	Aged or solution treated & aged		1.6
	200 - 425	Cast or cast & aged		1.6

Figure 99. Material Factors K (HP min / in^3) for Various Materials.

MATERIAL	BRINELL HARDNESS	DESCRIPTION	INDUSTRY	MATERIAL FACTOR K (HP min/in^3)
Titanium Alloys	400 MPa	Pure Ti	Used in very harsh environments that could cause considerable corrosion. Suitable for heat exchangers, desalting equipment, jet engine parts, landing gears & structural parts in aerospace	0.55
	950 MPa	α, near α & β alloys in annealed condition		0.65
	1050 Mpa	α + β alloys in aged condition, β alloys in annealed or aged condition (Ti6Al4V)		0.7
Inconel	32 - 40 Rc	Inconel 718	Rocket engines, & cryogenic applications	1.2
	< 220	Inconel 625	Marine & pertolium indusry for high strength & fracture toughness.	1.15
	< 195	Inconel 600	Variety of engine & airframe components which withstand high temperatures. Lockwire, exhaust liners, & turbine seals	1.15
Aluminum Alloys	30 - 80	Wrought & cold drawn	Engine block, cylinder head, transmission housings, casings, aerospace frame components.	0.3
	75 - 150	Wrought & solution treated & aged		0.3
	40 - 100	Cast		0.35
	70 - 120	Cast & solution treated & aged		0.35
	80	Unalloyed		0.15
	-	High Si Content 10 - 14%		0.35
	-	High Si Content 14 - 16%		0.55
Brass	55 - 61	Soft	Architectural fascias, grillwork, jewelry, ornamental trim, badges, door handles, marine hardware, primer caps.	0.35
	62 - 68	Medium	Watches, clock parts, hose fittings, plumbing valves, bicycle spoke nipples.	0.5
	69 - 73	Hard	Heat exchangers, containers, radiator cores, locks.	0.85
Bronze	40 - 165	Soft	Structural architecture, bearings, musical instruments, electrical contacts, ship propellers, machine tools.	0.35
	166 - 290	Medium		0.5
	291 - 415	Hard		0.85
Zinc	Vickers 30	Cast	Used to alloy with other metals, zinc galvanizing of parts, automobile electrical components, household fixtures, skin ointments, health benefits.	0.25

Figure 99. Continued.

14.1.2.2 IMPORTANCE OF POWER CALCULATION – CORRELATION TO MACHINE TOOLS

Power is perhaps the most common descriptor that is thrown around when it comes to machine performance, or specifically the machine tool's spindle capability. The Power draw of a machine spindle is one of the most visible gauges on most machine controllers. Actually, most often this gage reads a % Power draw (of the maximum Power available at that given RPM). Folks in the industry are often throwing around phrases such as: 'This machine of brand X is a 30 Hp machine'. Well, this is quite an inaccurate statement. As will be explained in a later section on 'Reading Power & Torque Curves', a machine's power is continually changing as we sweep through its range of RPMs.

A Rule-of-Thumb: In the machining of softer materials (i.e. Aluminum, Brass, Copper, etc.), Power is a significantly more important parameter in the selection of a machine tool, as compared to Torque. This is also true when it comes to employing High Velocity (shallow & fast) machining approach in hard metals. Such machine tools also tend to focus on higher RPM ranges. Typical modern-day machine tools used in soft metals and High Velocity machining are commonly available with max RPMs ranging from 15,000 to 30,000 rpm, and Power options ranging from 50 to 100 Hp. Such machines typically tend to have low Torque capability (~ less than 500 ft-lb) at these operating high RPMs.

14.2 CALCULATING TOTAL CUTTING FORCES IN METAL CUTTING

14.2.1 THE PHYSICS OF FORCES

In the world of physics, a force F is defined as:

$$Force = Mass \times Acceleration$$

14.2.2 CALCULATING TOTAL CUTTING FORCE IN METAL CUTTING

Visualize that the combination of workpiece material and the amount of cutter engaged, together pose a resistance to cutting. When the cutting tool approaches the workpiece at a pre-determined Speed & Feed, it will need to overcome this resistance and push through the cut. This ability to push through the cut can be summed up as the Total Cutting Force (F_T). In the machining world the Total Cutting Force (F_T) is calculated as:

$$F_T = \frac{HP \times 33{,}000}{SFM} \qquad [\text{Units: } lbs]$$

Where 33,000 is simply a conversion factor. An explanation of this conversion factor, as well as summary of this Total Cutting Force calculation can be seen in ***Figure 100***. In the interest of full disclosure, researchers can exhibit countless ways to not only calculate the Total Cutting Force, breakdown its vectors, and even analyze its variation at a cutting-edge progress through the arc of a cut. However, for the sake of this text the focus is kept a bit more macro, and shows a simplistic approach to calculating the Total Cutting Force. Although basic, this fundamental force calculation has proven to be a very practical analysis tool to researchers, and factory users alike. It can aid in process optimization, and even prevent catastrophic failures that result in stalling out of machines.

CUTTING FORCE (F_T) is the force reaction force exerted by the workpiece material onto the cutting tool during the cutting process. It is also known as the Total Cutting Force, or Tangential Force. Cutting force is typically measured in lb units.

$$F_T = \frac{HP \times 33{,}000}{SFM}$$

NOTE: **1HP = 550 ft lb / sec**
1HP = 550 x 60 ft lb / min
1HP = 33,000 ft lb / min

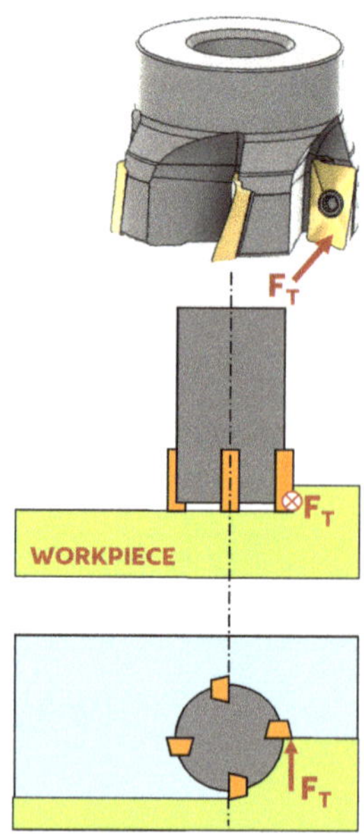

Figure 100. Calculating Total Cutting Force (F_t) in Milling.

14.2.2.1 IMPORTANCE OF FORCE CALCULATION – CORRELATION TO MACHINE TOOLS

At the most practical level, the lower the Total Cutting Force, the freer the cut, and consequently higher the machining efficiency. On the machine tool side, the Total Cutting Force is evaluated to confirm that we don't exceed the rating of the Axis Drives. Simply put, it should be recognized that each of the primary axis of a machine (X, Y, & Z) has an independent motor known as the Axis Drives. These Axis Drive motors are responsible for pushing the cutter through the cut.

In our everyday analogy of the lawn mower to a milling operation, the axis drive can be compared to the human being pushing the mower. Quite often while specking out a new machine, buyers will give a lot of attention to spindle performance, and significantly less (if at all) to the Axis Drive Ratings. Thus, making the choice of a very capable Spindle, and Axis Drive motors that become the weak link. When an Axis Drive motor is overloaded it spikes out to over 100% and can trigger stalling of the machine. This has been the cause of many a poor machine selection, and one that could be easily avoided by a Total Cutting Force calculation.

14.3 CALCULATING TORQUE IN METAL CUTTING

14.3.1 THE PHYSICS OF TORQUE

Is a force that tends to rotate an object. You generate a torque any time you apply force using a wrench. The object need not move when a torque is applied.

$$Torque = Force \times \ Distance\ from\ Center \qquad [Units:\ in.lb]$$

14.3.2 CALCULATING TORQUE IN METAL CUTTING

The Torque in metal cutting is calculated by multiplying the Cutting Force by the Radius of the cutter, ***Figure 101***. The equation is represented as:

$$TQ = F_T \times \left(\frac{D}{2}\right) \times \frac{1}{12} \qquad [\text{Units}:\ ft.lbs]$$

Typically, the metal cutting industry represents Torque in ft-lbs. Since the Diameter of the cutter is in inches, we need to do a unit's conversion by dividing it by 12 to obtain ft-lbs.

TORQUE (TQ) is simply the Cutting Force multiplied by the radius of the cutter. In metal cutting Torque is often measured in ft-lbf.

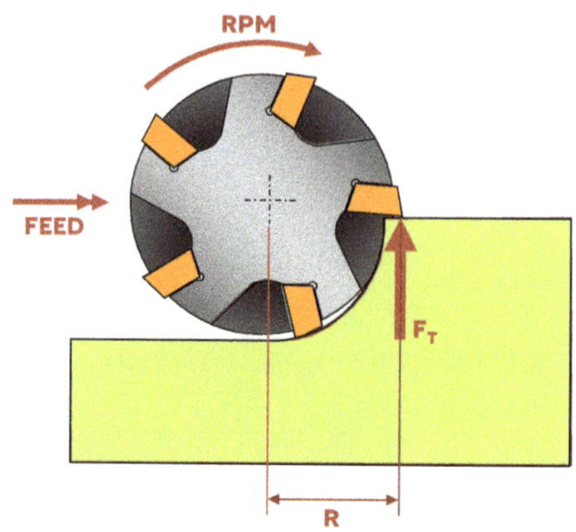

$$TQ = F_T \times R \text{ (in lbf)}$$

$$TQ = \frac{F_T \times R}{12} \text{ (in lbf)}$$

$$TQ = \left(\frac{HP \times 33{,}000}{SFM}\right) \times R$$

$$TQ = \left(\frac{HP \times 33{,}000}{\left(\frac{2\pi R(RPM)}{12}\right)}\right) \times R \times \frac{1}{12}$$

$$TQ = \frac{HP \times 5252}{RPM} \text{ (ftlbf)}$$

Figure 101. Torque Calculation.

14.3.2.1 IMPORTANCE OF TORQUE CALCULATION – CORRELATION TO MACHINE TOOLS

Torque is perhaps the least common descriptor, and arguably often overlooked parameter when it comes to selecting a machine tool spindle. In reality it is one of the most important parameters to keep an eye on. The Torque draw of a machine spindles is rarely visible as a gauge on machine controllers. If informed, folks in the industry are often throwing around phrases such as: 'This machine of brand X is a 900 ft.lb machine'. Well, this is quite an inaccurate statement. As will be explained in a later section on Reading Power & Torque Curves, a machine's Torque is continually changing as we sweep through its range of RPMs.

A Rule-of-Thumb: In the machining of harder materials (i.e. Steels, Titanium, Inconel, etc.) Torque is a significantly more important parameter in selecting of machine tools as compared to Power. Such machine tools also tend to focus on lower RPM ranges. Typical modern-day machine tools used in hard metals machining are commonly available with max RPMs ranging from 2,000 to 10,000 rpm, and Torque options ranging from 500 to 2500 ft.lb. Such machines typically tend to have low Power capability (~ less than 40 Hp) at these operating low RPMs.

14.4 READING POWER & TORQUE CURVES

As seem in ***Figure 1*** (TOP), the machine builder usually presents the spindle's power & torque curves by way of a logarithmic scale. This can make it very difficult to objectively interpret the data, and occasionally make the machine appear more favorable than it actually is. Data on the logarithmic scale can be skewed so that it appears as though the Power and Torque were peaking at the same time. It can also make certain RPM ranges look more favorable than others. The illustration at the bottom of ***Figure 102***, shows a power and torque curve cleaned up to a normal scale. Some may argue that some of the accuracy is lost in reformatting these curves to a normal scale. However, for the sake of this text we believe that the ability to clearly visualize and interpret the graph, far outweighs a marginal sacrifice in preciseness of the data.

Typically, a machine builders Power and Torque curves will show two lines for each:

- Continuous cycle line.
- Duty Cycle line.

The Continuous cycle line for Power and Torque is simply the maximum recommended values of operating under normal production conditions. If the Continuous cycle parameter were exceeded, the machine could still perform up to the Duty Cycle line, but for a short amount of time. This duty cycle can range from 10 mins to 30 mins, depending on the machine builder. It is always recommended that for normal production, the user stay below the Continuous cycle line.

As seen in ***Figure 102*** (Bottom), the RPM is shown on the X-Axis, the Power is shown on the primary Y-Axis (Blue Line), and the Torque is shown on the secondary Y-Axis (Pink Line). Additionally, from a safety standpoint, these curves only represent the Continuous Cycle lines of the machine. Typically, in any machine the Power initially climbs steadily and then plateau's out, while the Torque starts out at peak value and then tapers down.

Let's say we wanted to read the machine performance at 500 rpm. This is represented by the vertical dotted line. At 500 RPM, the machine has available Power = 110Hp (Blue, read on left Y-Axis), and Torque = 1500 ft-lb (Pink, read on right Y-Axis). At 2000 RPM, the machine has available Power = 110Hp (Blue, read on left Y-Axis), and Torque = 350 ft-lb (Pink, read on right Y-Axis).

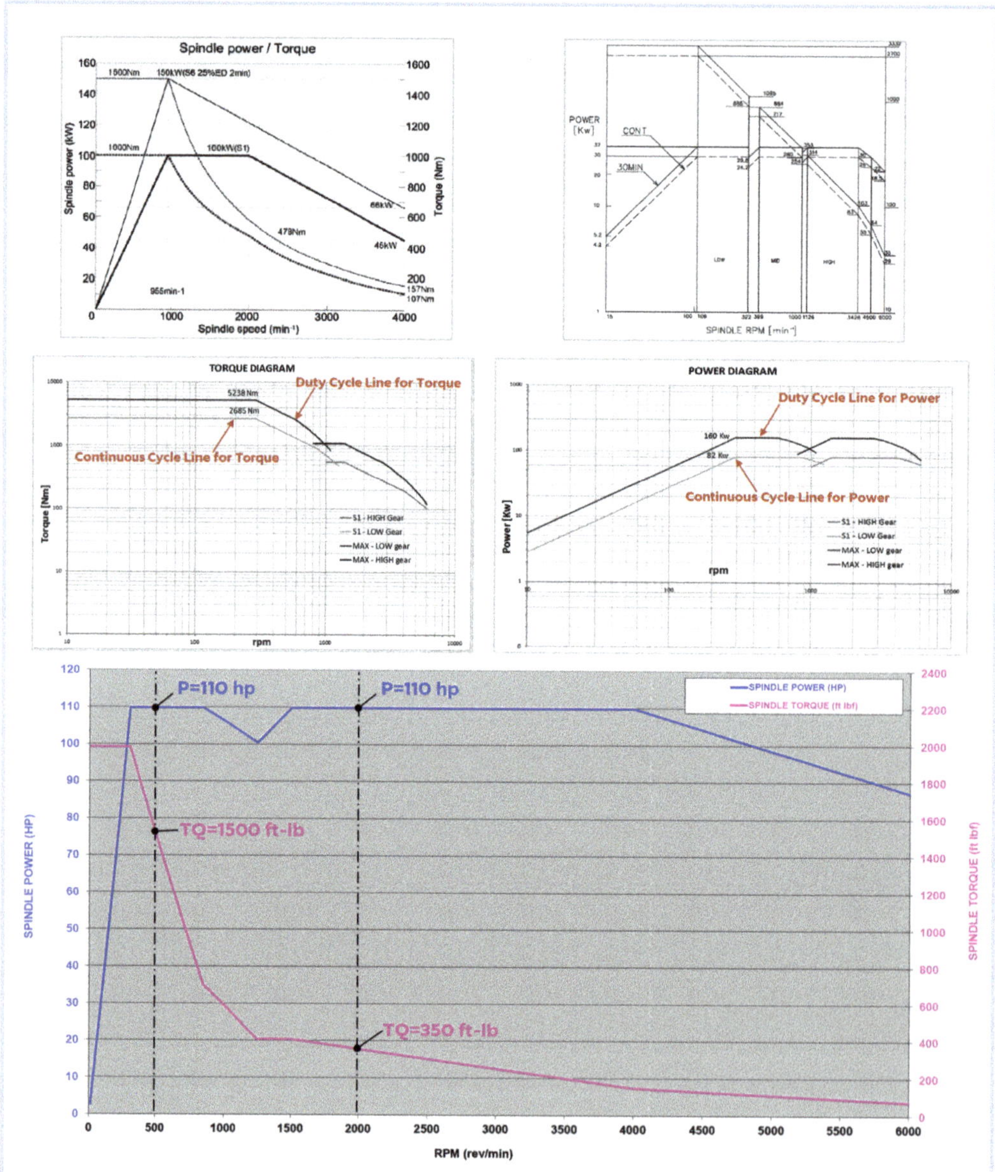

Figure 102. Reading HP & Torque Curves.

14.5 HOW MATERIAL FACTOR (K) TABLES ARE CREATED

Now that we have talked about the chain reaction of calculations: Power, Cutting Force, and Torque, it is important to backtrack and understand how Material Factor (K) tables are created.

14.5.1 MATERIAL FACTORS FROM DYNAMOMER ANALYSIS

This dynamometer approach to generate Material Factor is often employed in laboratory type environments. A dynamometer is a Sensor-Based equipment that is used to measure forces (***Figure 103***). The workpiece material to be cut is mounted on top of the dynamometer, and then is actually cut with a cutting tool at pre-determined speeds, feeds, and engagement parameters. As the cut is occurring, the dynamometer is continually feeding force data (X, Y, & Z) into a computer. These force vectors can then be converted into a Total Cutting Force (F_T). With a known F_T the Power can then be back calculated as:

$$Power = \frac{F_T \times SFM}{33{,}000} \qquad [Units: Hp]$$

It should also be realized that based on the cutting parameters, and engagement conditions, we also already know the Material Removal Rate:

$$MRR = DOC \times WOC \times IPM \qquad \left[Units: in^3 / \mathrm{min}\right]$$

The Material Factor (K) can now be figured out by:

$$K = \frac{POWER}{MRR} \qquad \left[Units: Hp.min / in^3\right]$$

Depending on the diligence of the laboratory running the experiment cutting parameters, engagement conditions, and cutter geometries would be tweaked prior to publishing a nominal Material Factor or its table.

CAUTIONARY NOTE: There are some consistent flaws is most published Material Factor tables, and can be summarized as:

- Most of these Force analysis experiments are done on a Lathe, and not a milling machine. Thus, the experiment is not taking into account any of the impact component of the cut
- The Force data is collected in front of the cutting tool and not from the back of the machine. Thus, if the intention of the calculations is to forecast machine

capabilities or limitations, this approach has not taken into account the efficiencies or inefficiencies of the machine (i.e. tool & holder interface type, gearing, resistance of moving parts, lubrication, mass of the system, vibrations, etc.).

NOTE: It is perfectly sound analysis if you were using the data collected by a dynamometer to: Develop a new cutter technology, compare between cuts for efficiencies, optimize cutter designs, optimize chip breaker geometries etc.

- Most of the published data often shows a singular value for K, when in reality the value should change and could be more accurately represented graphically.

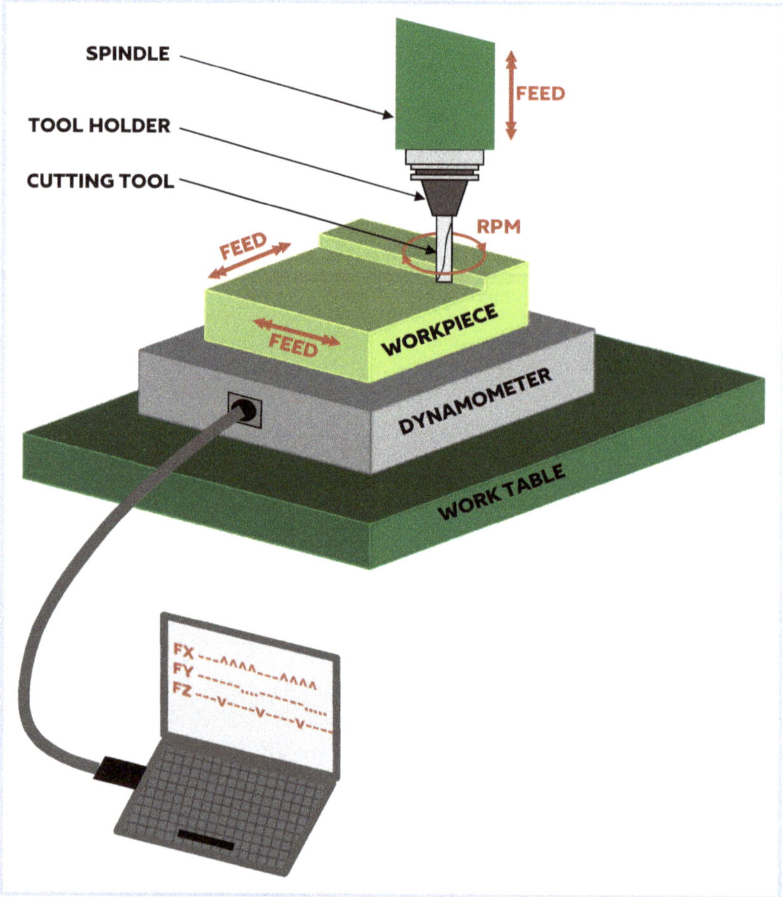

Figure 103. Material Factors from Dynamometer Analysis.

14.5.2 CREATING YOUR OWN MATERIAL FACTOR (K) LIBRARY ON PRODUCTION MACHINES

The technique discussed in this section could prove to be an invaluable tool for small and large metal cutting industry alike. Keep in mind that in a typical production environment one of the primary drivers in forecasting Power, Torque, and Cutting Force is to prevent bad things from happening to the machine-tool (i.e. crashes, stalls, catastrophic damage, & operator injuries). The predictive calculation of the above-mentioned parameters starts with using a reliable Material Factor (K).

Most machine tools come equipped with a controller panel with displays of Power Draw (%), ***Figure 104***. This reading in most parts is the Amperage draw by the spindle during cutting, thus taking into account all the system inefficiencies. For a cut of known cutting parameters, and engagement conditions (i.e. known MRR), the user could record the % Power draw from the machine controller. By referring to the machine's published Power & Torque curves the user could determine the exact Power being drawn (i.e. % of max power available at that RPM). Having done so, the Material Factor (K) can be arrived at by:

$$K = \frac{POWER}{MRR} \qquad \left[Units: \; Hp.min / in^3 \right]$$

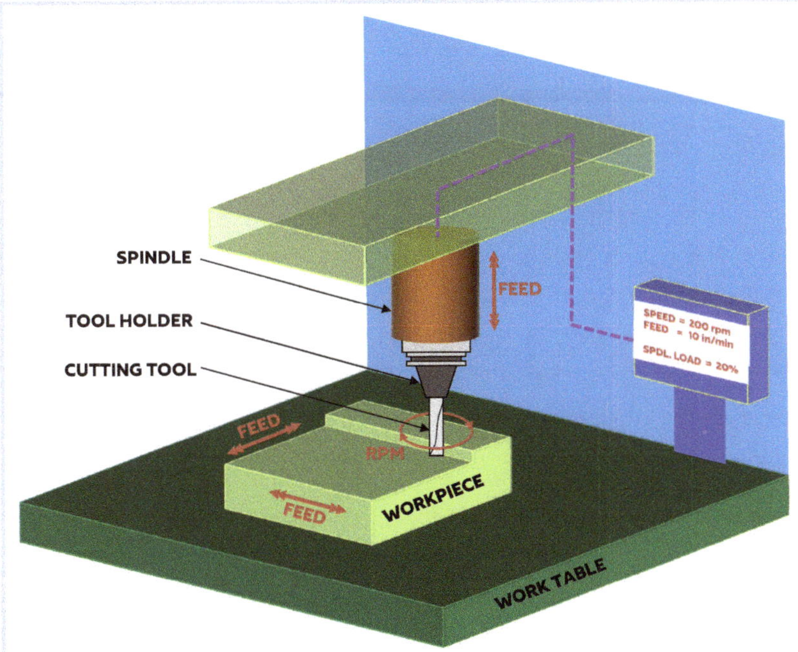

Figure 104. Material Factor from Production Machine Controller Data.

NOTE: This approach to creating a Material Factor K library is significantly more beneficial than simply the convenience of not having to use a dynamometer. Recognize that the machine controller is reading the spindle draw data from behind the motor (current Amp draw), and not from in front of the cutting tool. In doing so the controller data inherently incorporates all the inefficiencies of the machine (gearing, resistances, lubrication, bearings, spindle size, holder taper type & size etc.). Therefore, this data is more relevant towards predicting the behavior of future cuts in production. If the intent is to "prevent bad things from happening", then this approach is significantly relevant.

14.6 CALCULATING BENDING MOMENT IN METAL CUTTING

14.6.1 THE PHYSICS OF BENDING MOMENT

A bending moment is a force normally measured in a length x force (ft-lb). Bending moments occur when a force is applied at a given distance away from a point of reference; causing a "bending" effect (***Figure 105***, Top). In the simplest terms, a bending moment is basically a force that causes something to bend. If the object is not well-restrained the bending force will cause the object to rotate about a certain point.

This is often a difficult idea to comprehend, therefore consider an example of a plastic ruler overhanging a desk. If one end of the ruler is lying on the desk and is being held down, and then a force is applied to the other end of the ruler, it will cause the ruler to bend. The ruler will experience the greatest bending moment at the end where the force is applied.

To calculate the bending moment about a reference point A, we take the magnitude of the force and multiply it by the distance of the force from the point (***Figure 105***, Bottom). Often this is calculated across the span of a member. Note that the force must be perpendicular to the line between the point where the force is applied and the reference point. Though that sounds confusing, it can be simplified in our example:

It is important to note that to use the above formula, the force (in this case a 10 lb downward force) must NOT pass through the point. This is obvious as if we applied the same formula (distance x force) the distance would equal zero and therefore there would be no bending moment force.

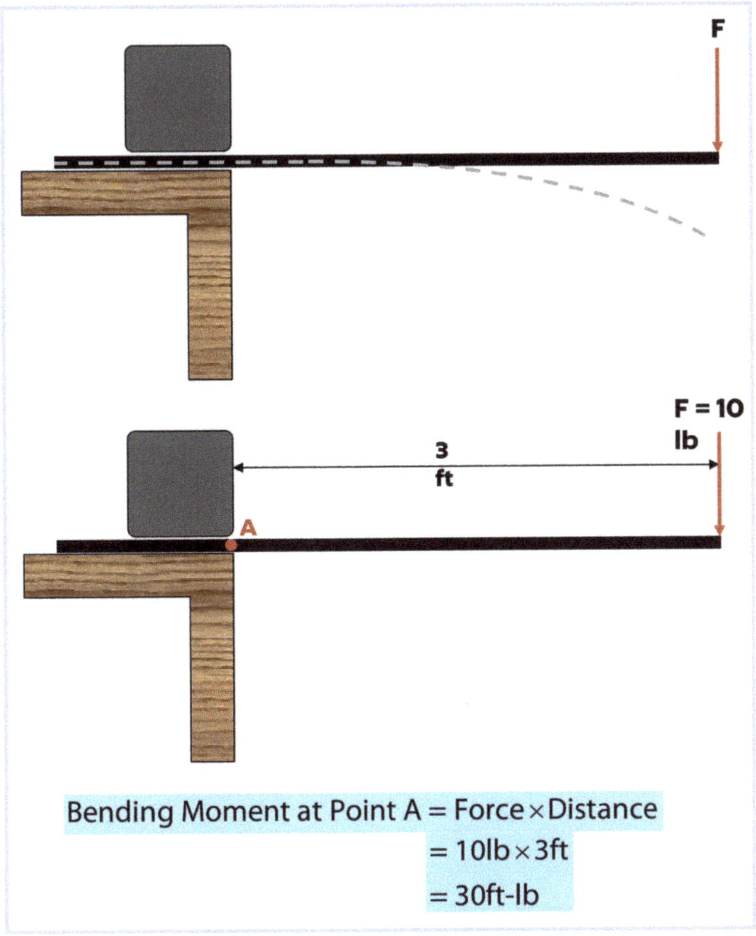

Figure 105. The Physics of Bending Moment.

14.6.2 CALCULATING BENDING MOMENT IN METAL CUTTING

The Bending Moment (B_M) in metal cutting is simply calculated by multiplying the Total Cutting Force (F_T) by the Gauge Length (G_L), ***Figure 106***.

$$B_M = F_T \times G_L \qquad [Units: in.lbs]$$

It should be noted that the metal cutting industry likes to view the Bending Moment in units of (in.lb), unlike Torque which is represented usually as (ft.lb).

BENDING MOMENT (B_m) is the moment created by the Total Cutting Force (F_T) at the location of the Gage Line. It is calculated by multiplying the Total Cutting Force by the Gage Length. Holder of various types & sizes have a published Bending Moment Limit. This is a necessary calculation to help confirm if the Total Cutting Force is or is not going to exceed its Bending Moment Limit, & cause separation of the holder-spindle interface. Bending Moment is typically measured in-lbs units.

$$B_m = F_T \times G_L$$

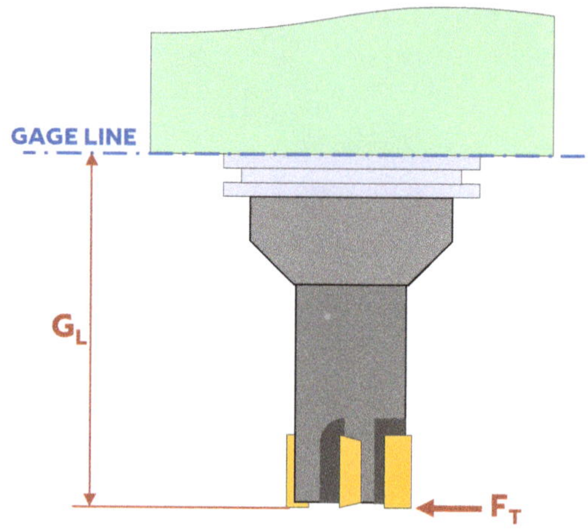

Figure 106. Calculating Bending Moment in Milling.

14.6.2.1 IMPORTANCE OF BENDING MOMENT CALCULATION - CORRELATION TO MACHINE TOOLS

The Bending Moment has a direct correlation towards selecting a Holder type and size for a machine. The part of a holder that goes into the spindle is known as the Taper. For each Taper type and size, are published maximum Bending Moment Limits. It is this maximum Bending Moment Limit (along with maximum RPM) that should be used in the selection of the ideal holder taper type and size. Consequently, this decision on holders, drives out the selection of the type of machine spindle needed. In depth explanation on holder taper type and size is discussed in subsequent chapter.

15

CONNECTING ALL THE DOTS

In the last few chapters we have focused on numerous metal-cutting concepts and mathematical computations. This chapter is strictly dedicated to tying it all together by actually doing the calculations, and visualizing the connections and chain reactions of variables. These calculations are further explained by special "Notes" to clarify the industry's reasoning on the numerical choices. For the sake of this analysis it is suggested that a certain level of trust be afforded in the validity of the input values as being industry standard. Running your own numbers and drawing conclusions is strongly recommended.

Imagine you are starting a factory with only two CNC milling machines. One machine is dedicated to machining Aluminum (a soft metal), and another for machining Titanium (a hard metal). Each of these machines are going to machine parts from forgings, and would probably invest 70% to 80% of the machining time in Roughing. Let's march through each of these scenarios with tried and true parameters that are typical in most modern-day industry. In the tabular storyboard shown in ***Table 4***, we compare Roughing in Aluminum and Titanium.

WORKPIECE		Aluminum	Titanium
OPERATION		Roughing	Roughing
NOTE: Ti will tap into the hardness & toughness attributes of carbide a lot more than Al. Therefore, Al cutter can accommodate significantly higher SFMs than Ti. This consequently ties into machine's RPM ranges for each material.			
CUTTER TYPE		Indexable Al Router	Indexable Rougher
DIA		= 2"	= 3"
# EFF		= 3	= 5
G_L		= 10"	= 10"
NOTE: The max RPM ranges required by machines for each material, also limit the size and shear mass of the cutting tools. After years of historical data and cost-benefit analysis the above chosen DIA offer the best bang of the buck.			
ENGAGEMENT CONDITIONS			
DOC	INPUT	= 0.75"	= 5.0"
WOC	INPUT	= 0.60"	= 1.0"
NOTE: The variation in DOC is simply a limitation driven by cutter mass driven by the need to tap into very high RPMs for Al, and lower RPMs for Ti. The WOC are simply calculated by the best practice of keeping it 30% of DIA.			
CUTTING PARAMETERS			
SFM	INPUT	= 10,000 ft/min	= 140 ft/min
RPM	= (SFM X 12)/(πDIA)	= 19,099 rev/min	= 178 rev/min
APT	INPUT	= 0.010"	= 0.005"
IPM	= APT X #EFF X RPM	= 573 in/min	= 4.5 in/min
CTM	$= DIA / 2[WOC (DIA - WOC)]^{1/2}$	= 1.09	= 1.06
IPM_{CTM}		Not Significant	Not Significant
NOTE: The input SFM and APT values are simply based on limitations of modern coated carbide cutting tools to yield a reasonable tool life. Think in terms of Hardness (friction & abrasive wear), & Toughness (impact & resistance to chipping).			
PERFORMANCE			
MRR	= DOC X WOC X IPM	= 258 in^3/min	= 22 in^3/min
NOTE: In the case of Al, despite having a smaller cutter & DOC, the ability to spin it a lot faster affords us a much higher MRR as compared to Ti. In the case of Ti, it is considered quite impressive to accomplish this MRR by modern industry standards.			
MACHINE LOADS			
K	INPUT	= 0.3 Hp.min/in^3	= 1.1 Hp.min/in^3
HP	= MRR X K	= 77 Hp	= 25 Hp

<table>
<tr><td colspan="3">NOTE: Notice how Al draws a significantly higher HP than Ti. This is one of the core reasons for the heavier focus on HP at high RPMs when selecting an Al machine.</td></tr>
<tr><td>F_T = (HP X 33000) / SFM</td><td>= 255 lbs</td><td>= 5777 lbs</td></tr>
<tr><td colspan="3">NOTE: Notice the significantly higher FT in Ti as compared to Al. This drives the Axis Drive Rating when selecting machine tools for each of the materials. The FT has direct impact on tool life. Also note how SFM finds its way back into the denominator of the FT calculation.</td></tr>
<tr><td>TQ = FT X (DIA/2) X (1/12)</td><td>= 21 ft.lbs</td><td>= 722 ft.lbs</td></tr>
<tr><td colspan="3">NOTE: Notice how Ti draws a significantly higher TQ than Al. This is one of the core reasons for the heavier focus on TQ at low RPMs when selecting a Ti machine.</td></tr>
<tr><td>B_M = FT X GL</td><td>= 2,552 in.lbs</td><td>= 57,773 in.lbs</td></tr>
<tr><td colspan="3">NOTE: As a direct consequence of the FT, the BM for Ti is significantly higher than that of Al. Every holder type and size has a published BM limit. Exceeding this limit will cause a separation between the holder and spindle face. This BM is the core driver in the selection of holders for the machining operation.</td></tr>
<tr><td>HOLDER TYPE</td><td>HSK63 & HSK80</td><td>CAT60 or larger
HSK125 or larger</td></tr>
<tr><td colspan="3">NOTE: It's not enough that the holder selection only falls within the published BM limit spec. Each holder type needs to also accommodate its max RPM allowable capability.</td></tr>
</table>

Table 4. A Comparison of Roughing in Al & Ti.

As we march down the chain reaction of computations, and review the notes on each of them, certain factors on differentiation between Aluminum (Al) and Titanium (Ti) machining become quite apparent:

- An Al (soft metals) machine is quite different from a Ti (hard metals) machine in terms of their performance bandwidths. If these machines were used interchangeably in a production environment, the yield would be greatly diminished in terms of efficiency, cycle times, wear & tear, and ultimately costs.
- The RPM range in an Al machine is significantly higher than that in a Ti machine.
- The Al machine requires in its performance band significantly higher Power as compared to the Ti machine.
- The Ti machine requires in its performance band, significantly higher Torque as compared to the Al machine.
- The Cutting Forces in a Ti machine are significantly higher than those in an Al machine and therefore will affect the ratings of their respective Axis Drives.

- As a function of the Cutting Forces and Gage Lengths (assume same in both), the Bending Moment in Ti is significantly higher in Ti than in Al machine. This consequently affects the choices of tool holding systems. Tool holder taper type & size should be selected based on their published Bending Moment limits.

Now that we have a strong understanding of how the two materials affect output parameters on the machine side, let's run an analysis for Finishing in each of the materials. ***Table 5*** and ***Table 6*** show side-by-side comparison of Roughing and Finishing parameters in Al and Ti respectively.

WORKPIECE		Aluminum	Aluminum
OPERATION		Roughing (From ***Table 4***)	Finishing
CUTTER TYPE		Indexable Al Router	Solid Carbide Endmill
DIA		= 2"	= 1"
# EFF		= 3	= 5
G_L		= 10"	= 10"
NOTE: It is recommended to use Solid Carbide Endmills in Finishing operations, as they naturally hold a tighter tolerance over an Indexable cutter.			
ENGAGEMENT CONDITIONS			
DOC	INPUT	= 0.75"	= 2.0"
WOC	INPUT	= 0.60"	= 0.05"
NOTE: In Finishing it is recommended to engage as much of the Flute Length of the Solid Carbide Endmill as possible. This allows for the best Surface Finish. Also, in finishing the WOC is kept relatively lighter as a safety net for finish size and tolerances.			
CUTTING PARAMETERS			
SFM	INPUT	= 10,000 ft/min	= 5000 ft/min
RPM	= (SFM X 12)/(πDIA)	= 19,099 rev/min	= 19,099 rev/min
APT	INPUT	= 0.010"	= 0.004"
IPM	= APT X #EFF X RPM	= 573 in/min	= 4.5 in/min
CTM	= DIA / 2[WOC (DIA – WOC)]$^{1/2}$	= 1.09	= 2.29
IPM_{CTM}	= IPM X CTM	Not Significant	= 1051 in/min
NOTE: In Finishing the SFM is dropped so as to accommodate the max RPM of a typical modern-day machine. Also, the APT is kept relatively lighter so as not to create Feedrate marks by way of scallops on the final part finish. Notice that the Chip Thinning Multiplier (CTM) in Finishing is a healthy 2.29, and worthy of factoring in with for the IPM to yield IPMCTM.			

PERFORMANCE		
MRR *Roughing* = DOC X WOC X IPM *Finishing* = DOC X WOC X IPM_{CTM}	= 258 in^3/min	= 105 in^3/min
NOTE: The Roughing MRR factors in IPM without the CTM, while Finishing MRR factors in the IPM with CTM.		
MACHINE LOADS		
K INPUT	= 0.3 Hp.min/in^3	= 0.3 Hp.min/in^3
HP = MRR X K	= 77 Hp	= 32 Hp
F_T = (HP X 33000) / SFM	= 255 lbs	= 208 lbs
NOTE: Despite the HP in Roughing being almost double that of Finishing, the Cutting Forces FT are relatively closer to each other. This is primarily driven out by the fact that the SFM play a role in the FT calculation. The SFM in Finishing is half that in Roughing.		
TQ = FT X (DIA/2) X (1/12)	= 21 ft.lbs	= 8.7 ft.lbs
B_M = FT X GL	= 2,552 in.lbs	= 2082 in.lbs
HOLDER TYPE	HSK63 & HSK80	HSK63 & HSK80
NOTE: BM is a product of GL and FT. In this case the machine can only have one holding system, and is predominantly governed by the Roughing parameters.		

Table 5. Comparing Roughing & Finishing Parameters in Aluminum Machining.

WORKPIECE		Titanium	Titanium
OPERATION		Roughing	Finishing
CUTTER TYPE		Indexable Rougher	Solid Carbide Endmill
DIA		= 3"	= 1"
# EFF		= 5	= 5
G_L		= 10"	= 10"
NOTE: It is recommended to use Solid Carbide Endmills in Finishing operations, as they naturally hold a tighter tolerance over an Indexable cutter.			
ENGAGEMENT CONDITIONS			
DOC	INPUT	= 5.0"	= 2"
WOC	INPUT	= 1.0"	= 0.05"
NOTE: In Finishing it is recommended to engage as much of the Flute Length of the Solid Carbide Endmill as possible. This allows for the best Surface Finish. Also, in finishing the WOC is kept relatively lighter as a safety net for finish size and tolerances.			
CUTTING PARAMETERS			
SFM	INPUT	= 140 ft/min	= 300
RPM	= (SFM X 12)/(πDIA)	= 178 rev/min	= 1146 rev/min
APT	INPUT	= 0.005"	= 0.005"
IPM	= APT X #EFF X RPM	= 4.5 in/min	= 28.6 in/min
CTM	= DIA / 2[WOC (DIA – WOC)]$^{1/2}$	= 1.06	= 2.29
IPM_{CTM}		Not Significant	= 65.7 in/min
NOTE: In Finishing the SFM is dropped so as to accommodate the max RPM of a typical modern-day machine. Also, the APT is kept relatively lighter so as not to create Feedrate marks by way of scallops on the final part finish. Notice that the Chip Thinning Multiplier (CTM) in Finishing is a healthy 2.29, and worthy of factoring in with for the IPM to yield IPMCTM.			
PERFORMANCE			
MRR	= DOC X WOC X IPM	= 22 in^3/min	= 6.6 in^3/min
NOTE: The Roughing MRR factors in IPM without the CTM, while Finishing MRR factors in the IPM with CTM.			
MACHINE LOADS			
K	INPUT	= 1.1 Hp.min/in^3	= 1.1 Hp.min/in^3
HP	= MRR X K	= 25 Hp	= 7.2 Hp
F_T	= (HP X 33000) / SFM	= 5777 lbs	= 795 lbs

NOTE: Notice how in machining of Ti the HP draw is significantly lower than in the previously discussed Al machining. In Ti machining the FT is significantly higher than that in Al. Also, it should be recognized that the SFM plays a factor in the FT calculation.			
TQ	= FT X (DIA/2) X (1/12)	= 722 ft.lbs	= 33 ft.lbs
NOTE: Notice how in machining of Ti the TQ draw is significantly higher than in the previously discussed Al machining.			
B_M	= FT X GL	= 57,773 in.lbs	= 7953 in.lbs
HOLDER TYPE		CAT60 or larger HSK125 or larger	CAT60 or larger HSK125 or larger
NOTE: BM is a product of GL and FT. In this case the machine can only have one holding system, and is predominantly governed by the Roughing parameters.			

Table 6. Comparing Roughing & Finishing Parameters in Titanium Machining.

Recognizing that the performance spec for machine-tool to cut Aluminum is quite varied from that to cut Titanium, we will now attempt to select machine parameters. Based on just the input - output parameters of just the two cutters (Roughing & Finishing) engaged in Aluminum and Titanium machining, let's try to create an ideal Power & Torque Curve for each of the required machines (***Figure 107*** and ***Figure 108***). Additionally, for the sake of visibility ***Figure 108*** also shows a zoomed illustration where the max RPM is capped off at 1500 rpm.

Also shown on each of these curves (***Figure 107*** and ***Figure 108***) are the dots representing the HP & Torque draw values for Roughing and Finishing. These are the values of HP & Torque calculated earlier in the tables shown in ***Table 5*** and ***Table 6***. For the machine to execute these cuts, the HP dots must fall well below the HP curve line, and the Torque dots must fall well below the Torque curve line.

In closing, to truly appreciate how varied the performance band of an Aluminum machine is from that of a Titanium machine, ***Figure 109*** shows each of their curves drawn to the same scale on the X & Y axis. These HP & Torque curves could aid in starting the conversation with a machine builder in terms of baseline requirements for the machine spindles.

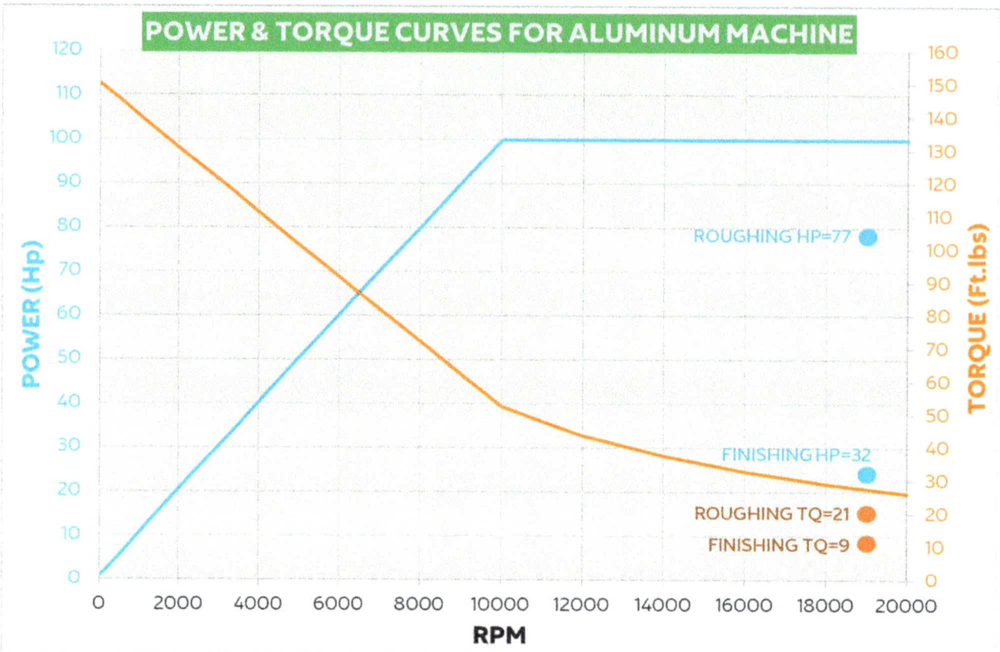

Figure 107. Generic HP & Torque Curves for Workshop on Aluminum Machine.

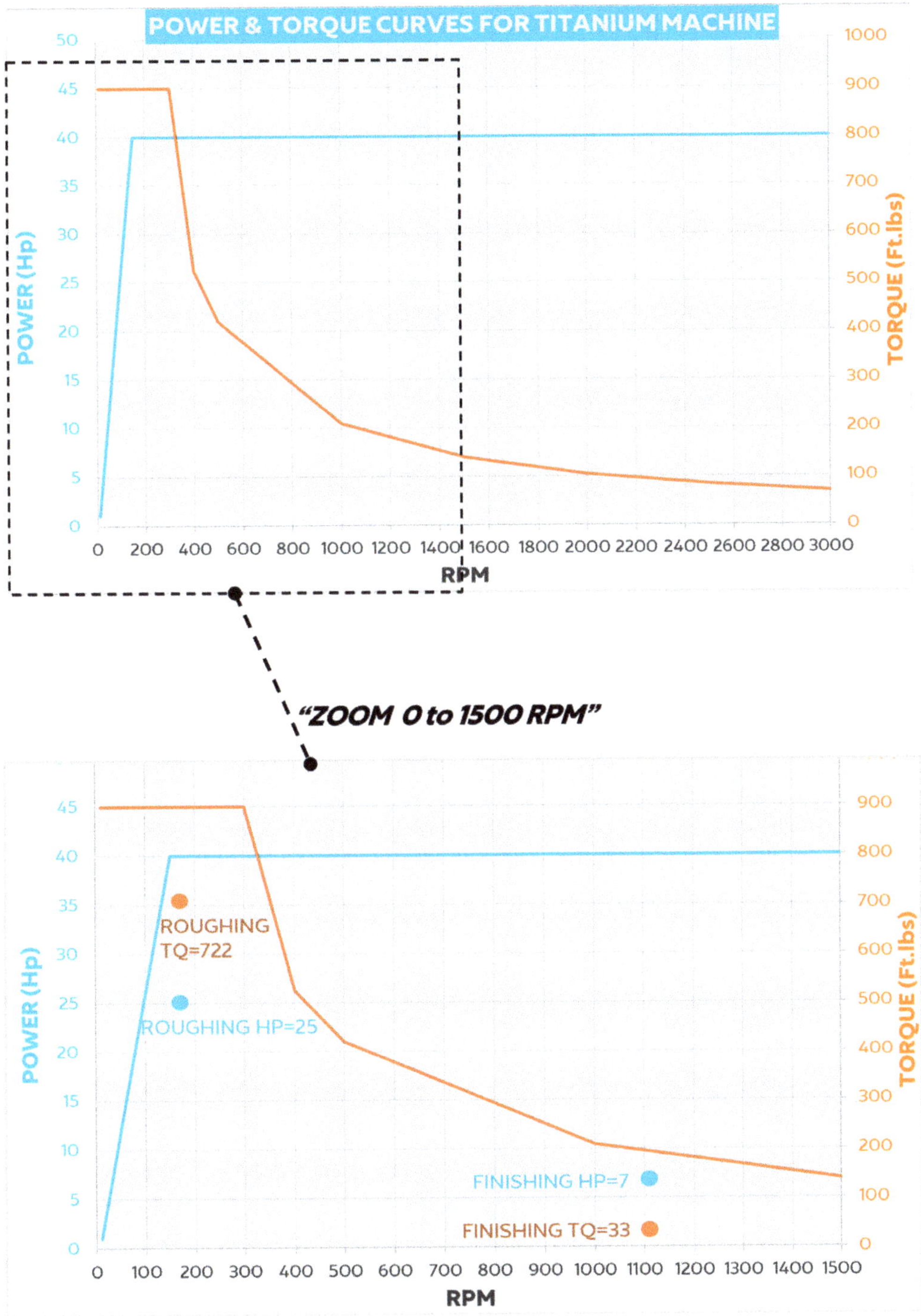

Figure 108. Generic HP & Torque Curves for Workshop on Titanium Machine.

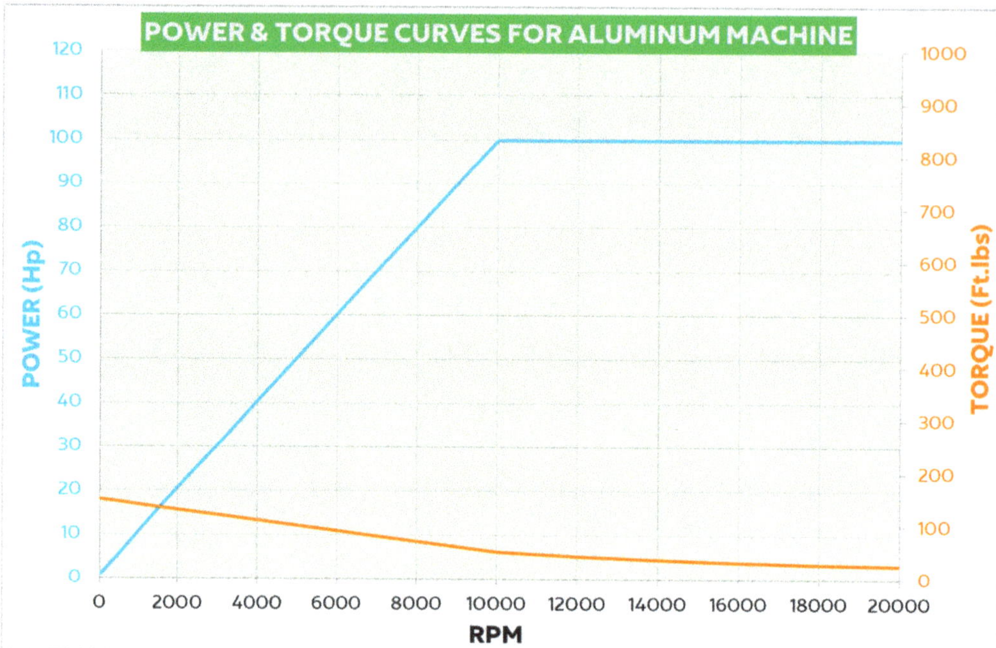

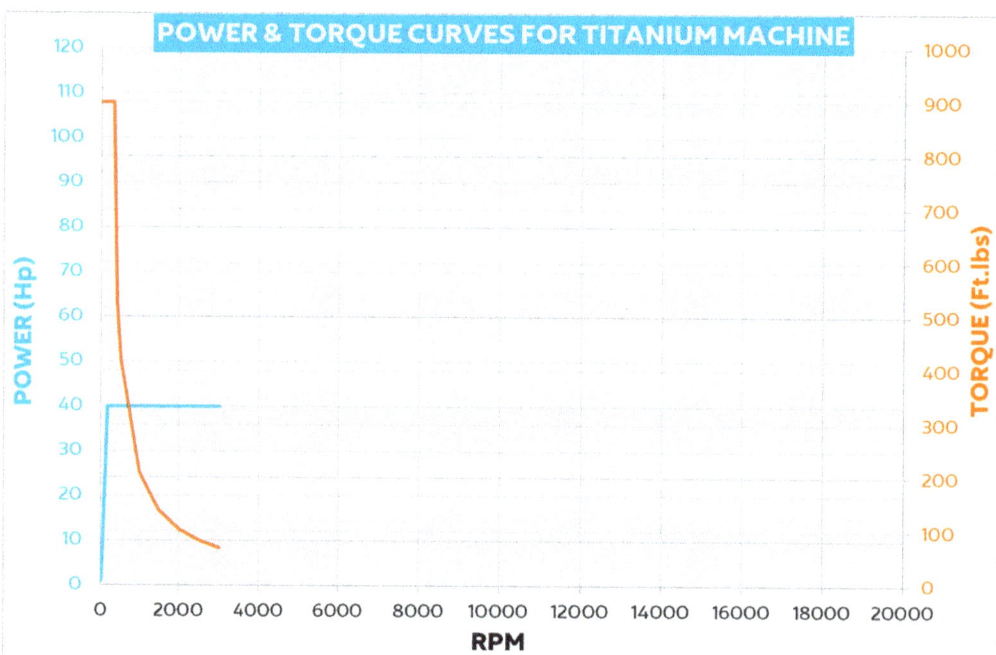

Figure 109. Generic HP & Torque Curves for Al & Ti Machines [At same scale on X & Y Axis].

16

TOOL & HOLDER BALANCING

The concept of tool & holder balancing is an integral part of a best practice when it comes to machining materials at high RPMs. More critically, the user should have a firm understanding of not only when balancing is necessary, but also when it can be detrimental (and/or not repeatable). This chapter not only deep dives into the core theory of tool & holder balancing, but also explains how to use data-driven decisions to justify not balancing certain systems.

Interestingly, to fully grasp the theory of balancing, we will actually spend a lot time and energy towards the understanding of "Unbalance". This sometimes-counterintuitive concept of unbalance, is one of the primary drivers in the mathematics of balance calculations.

16.1 WHAT IS UNBALANCE

The concept of unbalance may be naturally intuitive to many end-users, and can be further matured by way of some visualizing techniques as well as math. Let's start by visualizing a simple large cylinder made of a relatively soft material (say plastic), ***Figure 110***. Next, visualize a smaller cylinder of a denser material (say steel) is pressed into the larger plastic cylinder. For the sake of the math involved let's say that the steel cylinder is located at a distance R (mm) from the center of the plastic cylinder, and that it has a mass g (grams). Therefore, the unbalance caused to the plastic cylinder by the implanted steel cylinder can be calculated as:

$$UNBALANCE = Mass\,(grams) \times Distance\,(mm)$$

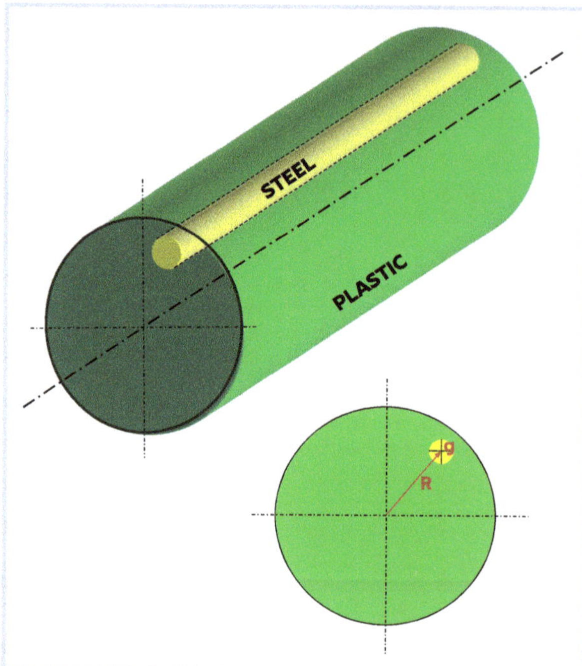

Figure 110. Understanding Unbalance.

Example: If the mass of the implanted steel cylinder were 2 g, and it was located 12 mm from the center of the plastic cylinder, then the total unbalance would be calculated as 24 g-mm (***Figure 111***).

[Let's try to put some everyday perspective on these values, & how small changes can have big impact on unbalance. A single dollar bill weighs about 1 g. A stack of about 10 sheets of everyday use printer paper measures about 1 mm.]

Now back to our cylinder example: Lets change the mass of the steel cylinder by just one gram to 3 g, and its location by one millimeter to 13mm, (***Figure 112***). Now the unbalance of the system changes to 39 g-mm. These minute changes work out to a 62% increase in unbalance.

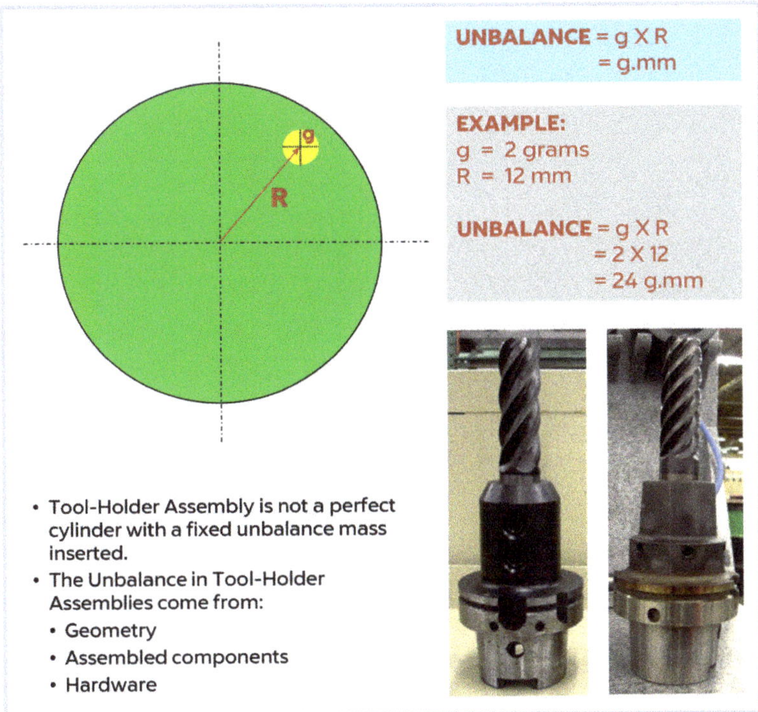

Figure 111. Understanding Unbalance – Basic Calculations.

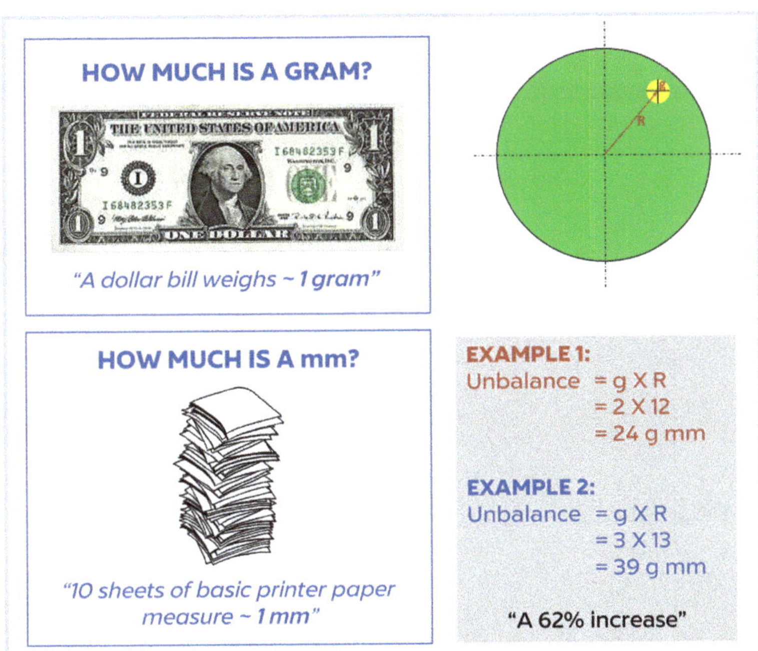

Figure 112. Understanding Unbalance – Impact of Small Changes.

16.2 ACCEPTABLE UNBALANCE

The concept of acceptable unbalance is the primary driver behind the critical calculations needed to balance a system. The Acceptable Unbalance U (g mm) can be viewed as the maximum allowable unbalance that a too-holder assembly can accommodate. The greater the U, the greater the Acceptable Unbalance. Another way to visualize Acceptable Unbalance is to ask yourself the question, "How much unbalance can I live with?". Greater the value of U, means greater the ability of the tool-holder assembly to accommodate the forces created by the unbalance F_U. The Acceptable Unbalance U can be calculated by the equation below.

$$Acceptable\ Unbalance\ U(g\ mm) = \frac{G \times 9549 \times Kg}{RPM}$$

Where G represents the G-Scale for required level of balancing precision, Kg is the weight in kilograms of the tool-holder assembly, and RPM is the rotational speed of the assembly. The influence of each of these variables will be explored in detail in the subsequent sections.

16.2.1 INFLUENCE OF RPM

As shown in ***Figure 113***, a long slender cylinder is made to spin at a relatively low speed of 10 rpm. Now, let's spin this same cylinder at a significantly higher speed of 10,000 rpm. Intuitively we can visualize that the cylinder will want to deflect at the higher RPM due to centrifugal forces, as compared to the lower RPM. Now let's connect this intuition with the mathematics of Acceptable Unbalance U.

Since RPM is in the denominator of the Acceptable Unbalance U equation; as the RPM goes up, the Acceptable Unbalance U goes down. Another more practical way to visualize this is to recognize that as the system spins faster, the amount of unbalance we can live with gets smaller. Therefore, relating this concept to machining, it is very important to recognize either:

- The maximum RPM at which the tool-holder assembly will operating.

OR

- The maximum RPM of the machine.

These values are key input parameters needed in a balancing machine.

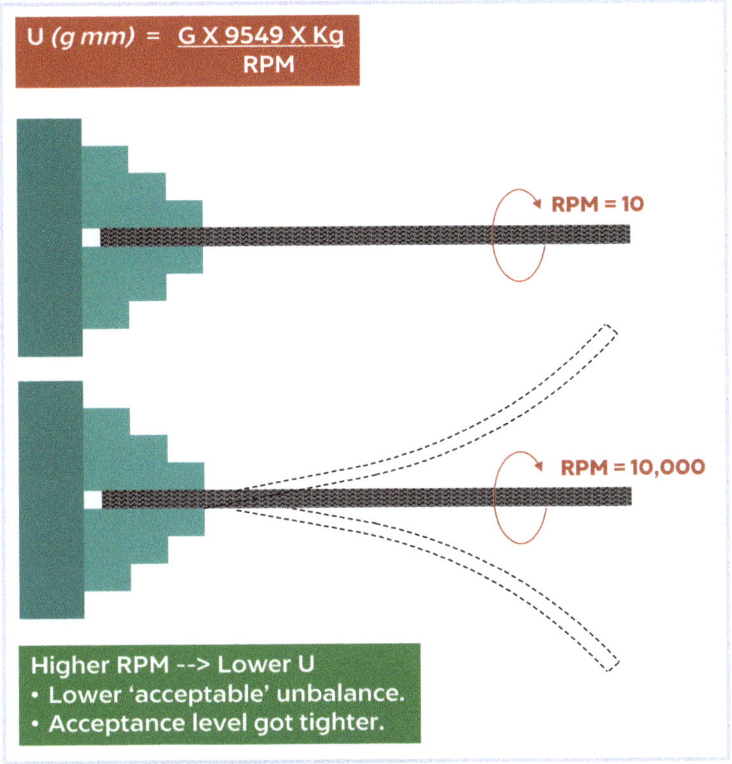

Figure 113. Understanding Unbalance – Influence of RPM.

16.2.2 INFLUENCE OF WEIGHT

Now as shown in ***Figure 114*** lets visualize the starting slender cylinder weighing 1kg and spinning at 10,000 rpm, is replaced with a beefier 12 kg cylinder. Intuitively we recognize that the cylinder with the now new larger mass will stabilize from the previously observed deflection. Increasing the weight of the system made the Acceptable Unbalance U higher. In other words, by increasing the weight we are able to live with a lot more unbalance without detrimentally affecting the system.

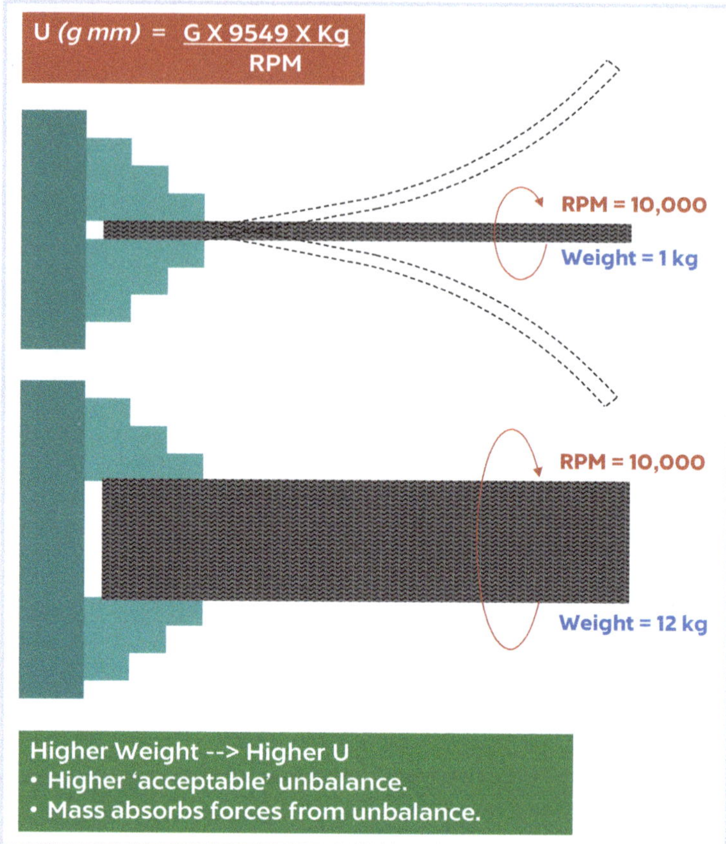

Figure 114. Understanding Unbalance – Influence of Weight (Kg).

16.2.3 LEVEL OF PRECISION – THE G-SCALE

The level of precision required in the Acceptable Unbalance U equation is established by the value of the G-scale. In selecting the G value some noteworthy questions to ask are:

- How precise do I need it?
- Is that level of precision necessary?
- Can we physically accomplish that level of precision?
- Can we repeat the level of precision established?

For Example: If you were asked to measure the length & width of a table so that you could purchase a tablecloth to cover it with, what measurement technique would be precise enough? As shown in ***Figure 115*** the options for such a measurement could range from: an eyeball estimate, a ruler or tape measure, dial calipers, micrometer, a CMM machine, or palm with measurements. It becomes quite apparent that while a ruler or tape measure with do the job sufficiently, a CMM machine would simply be ridiculous. A similar thought process is applied in selecting the G value. Let's walk through a few more examples to further enhance our understanding of established G values.

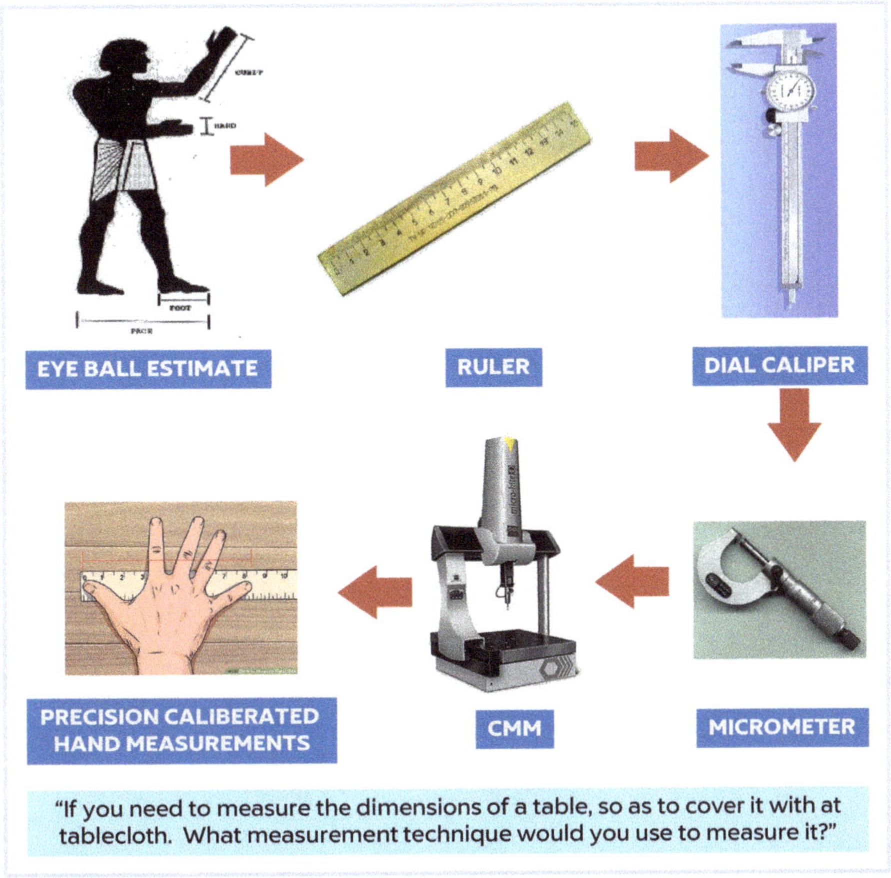

Figure 115. Understanding Unbalance – The G-Scale Levels of Precision.

Typically, the G value is selected from a pre-established table (ISO-1940/1, ANSI S2.19), ***Figure 116***. These values are baseline values arrived at experimentally so as to create reasonable, repeatable input parameter for calculating Acceptable Unbalance & consequently balancing.

ISO-1940

$$U\ (g\ mm) = \frac{G \times 9549 \times Kg}{RPM}$$

Machinery types: General examples	Balance quality grade G	Magnitude $e_{per} \cdot \Omega$ mm/s
Crankshaft drives for large slow marine diesel engines (piston speed below 9 m/s), inherently unbalanced	G 4000	4 000
Crankshaft drives for large slow marine diesel engines (piston speed below 9 m/s), inherently balanced	G 1600	1 600
Crankshaft drives, inherently unbalanced, elastically mounted	G 630	630
Crankshaft drives, inherently unbalanced, rigidly mounted	G 250	250
Complete reciprocating engines for cars, trucks and locomotives	G 100	100
Cars: wheels, wheel rims, wheel sets, drive shafts Crankshaft drives, inherently balanced, elastically mounted	G 40	40
Agricultural machinery Crankshaft drives, inherently balanced, rigidly mounted Crushing machines Drive shafts (cardan shafts, propeller shafts)	G 16	16
Aircraft gas turbines Centrifuges (separators, decanters) Electric motors and generators (of at least 80 mm shaft height), of maximum rated speeds up to 950 r/min Electric motors of shaft heights smaller than 80 mm Fans Gears Machinery, general Machine-tools Paper machines Process plant machines Pumps Turbo-chargers Water turbines	G 6,3	6,3
Compressors Computer drives Electric motors and generators (of at least 80 mm shaft height), of maximum rated speeds above 950 r/min Gas turbines and steam turbines Machine-tool drives Textile machines	G 2,5	2,5
Audio and video drives Grinding machine drives	G 1	1
Gyroscopes Spindles and drives of high-precision systems	G 0,4	0,4

Industry Standard for High Speed Machining (> 6000 RPM) with Solid Carbide Endmills in Shrink Fit Holders

Figure 116. The G-Scale ISO & ANSI Standard.

A Ferris wheel (***Figure 117***) has a G value that is so high it is futile to attempt balancing it. Additionally, a Ferris wheel has multiple large components that are assembled at variable tolerances, each contributing to an unbalance condition. The combination of the mass being very large, and the RPMs low makes the Acceptable Unbalance U large. Thus, making a Ferris wheel virtually impossible & unnecessary to balance.

A car tire & wheel assembly (***Figure 117***) typically has a G value of 40. An average car driving at 100 mph would be spinning at 1500 rpm. Considering that a car tire assembly has fewer components that a Ferris wheel, some variability in air pressure, and torque of tightening bolts, its acceptable Unbalance U is calculated with a reasonable G value of 40.

A grinding wheel assembly or electronic motor needs to be balanced significantly more precisely than a car tire assembly (***Figure 118***). In these instruments we can live with very little Acceptable Unbalance U. Consequently, the G value of grinder is set at 6.3 and for a small electronics motor at 0.4.

Figure 117. Unbalance - G-Scale & How Precise do you need to be?

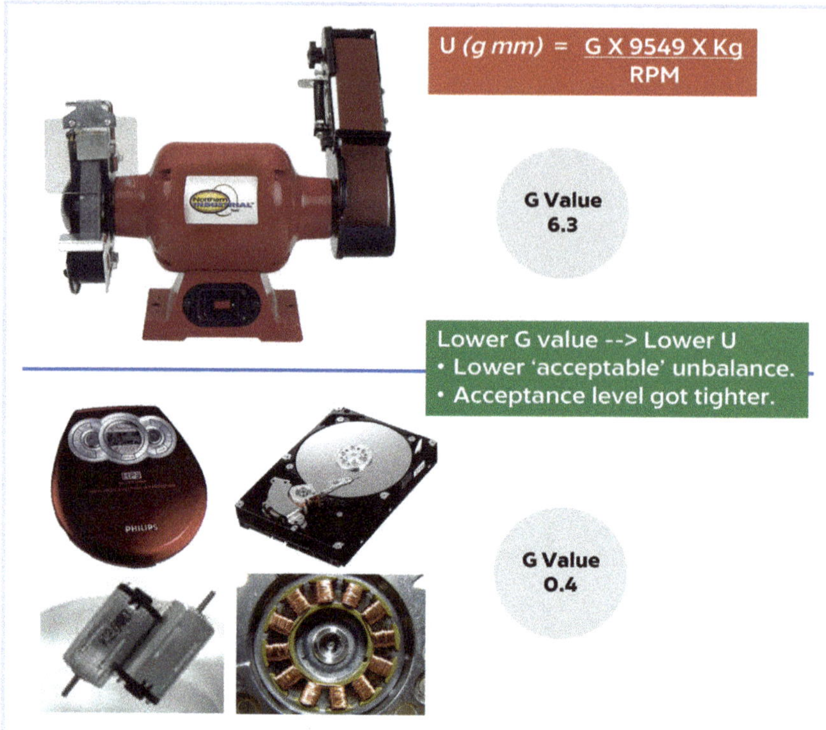

Figure 118. Unbalance - G-Scale & How Precise do you need to be? (contd.)

16.2.4 THE 9549 CONVERSION FACTOR

For those of you who just won't sleep well until the unknow if fully resolved, the value of 9549 in the Acceptable Unbalance U equation in the simply a conversion factor. As shown in ***Figure 119***:

- The units of U are (g.mm).
- The units of G are (mm/sec).
- The weight is represented in kilograms (kg).
- The units of speed are (revolutions/min).

The time units need to be converted to seconds, the weight units need to be converted to grams, and the revolutions need to be converted to radians. Therefore:

$$\frac{60 \times 1000}{2\pi} = 9459.2966$$

16.2.5 UNBALANCE - PUTTING IT ALL TOGETHER

Before we proceed talk about how a balancing machine works, let's put all the input parameters to work by way of a practical example. This understanding will help visualize the logic of a balancing machine. ***Figure 119*** shows all the key parameter (G, Kg, & RPM) used towards calculating the Maximum Allowable Unbalance U. These are also the input parameters that a balancing machine needs for its computations.

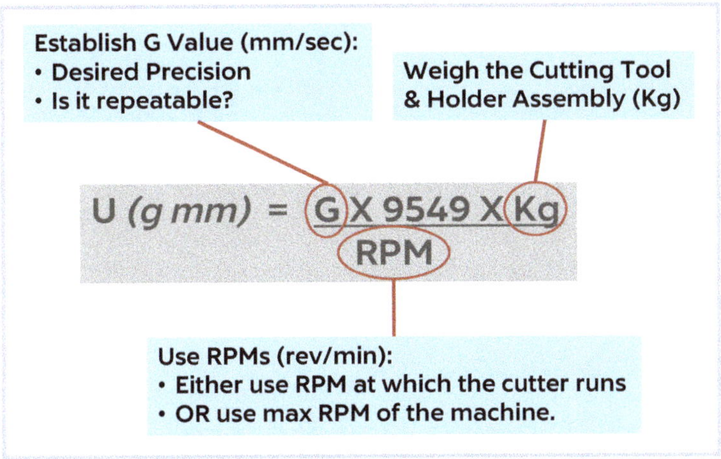

Figure 119. Understanding Unbalance – Input Parameters for a Balancing Machine.

The example shown in ***Figure 120*** walks through how small changes in input parameters can affect the value of U.

- It starts with G=1, Weight=2.75kg, RPM=8000. This works out to **U=3.3 g mm**.
- Next the user realizes that the G value is too small, and changes it to G=2.5. This works out to **U=8.2 g mm**.
- Next the user prefers to change the RPM to 16,000 (to be the max rpm of the machine). This works out to **U=4.1 g mm**.
- Finally, the user realizes that they did not weigh the tool earlier, and correct the weight to 8.25 kg. This works out to **U=12.3 g mm**.

As is evident, small changes in input parameters make a significant change in the Acceptable Unbalance U. Considering from start to end, with all the small corrections in input, the U value changed from 3.3 to 12.3 g mm. This is a 372% increase. It is important to recognize that a balancing machine is going to be only as effective as the correctness of its input parameters. The machine will always attempt to give a balancing recommendation. However, it the input parameters are not correct or reasonable, the balancing will either not show in performance, or it may not be repeatable.

$$U\ (g\ mm) = \frac{G \times 9549 \times Kg}{RPM}$$

$$U_{1.0} = \frac{1.0 \times 9549 \times 2.75}{8000} = 3.3\ g\ mm$$

Observe the impact of these variables

Increase G value

$$U_{2.5} = \frac{2.5 \times 9549 \times 275}{8000} = 8.2\ g\ mm$$

Increase G value
Now we can live with more Unbalance

Double the RPM

$$U_{2.5} = \frac{2.5 \times 9{,}549 \times 2.75}{16{,}000} = 4.1\ g\ mm$$

Increase RPM
Now we can live with less Unbalance

Triple the Weight

$$U_{2.5} = \frac{2.5 \times 9{,}549 \times 8.25}{16{,}000} = 12.3\ g\ mm$$

Increase Weight
Now we can live with more Unbalance

Figure 120. Calculating Max Allowable Unbalance.

16.3 THE BALANCING MACHINE

Figure 121 show a typical balancing machine, and its Input/Output logic. As discussed earlier, the balancing machine needs the user to correctly input values for G, Weight, and RPM. It then calculates the Acceptable Unbalance U. Next, the machine spins the cutting tool & holder assembly. By way of sensors, it is then able to establish the Unbalance Forces that are at play in the system. It then compares this against the calculated value of U, and makes a recommendation to counteract these Unbalance Forces. The recommendation is by way of counterbalance theory, ***Figure 122***. It identifies the exact location where, and amount of mass that needs to be added or subtracted to the cutting tool & holder assembly. As shown in in ***Figure 122*** there are available holder termed 'Balanceable'. They come equipped with pre-drilled & tapped holes to accommodate weighted setscrews. This is also very much the same approach to balancing of car tires.

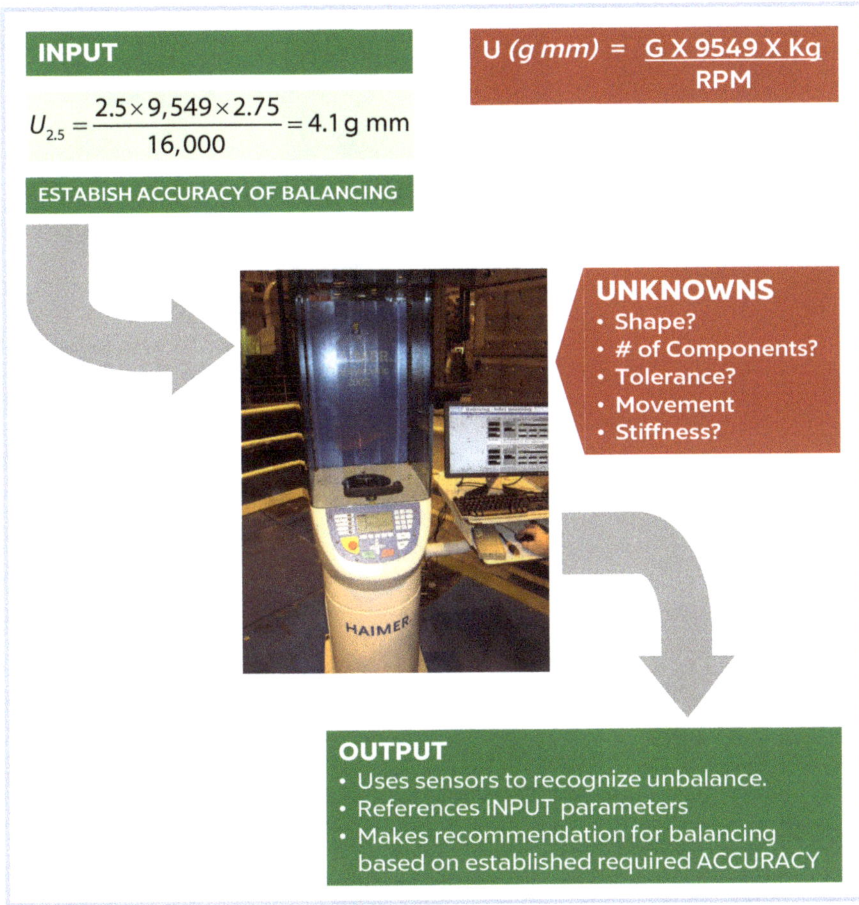

Figure 121. The Balancing Machine.

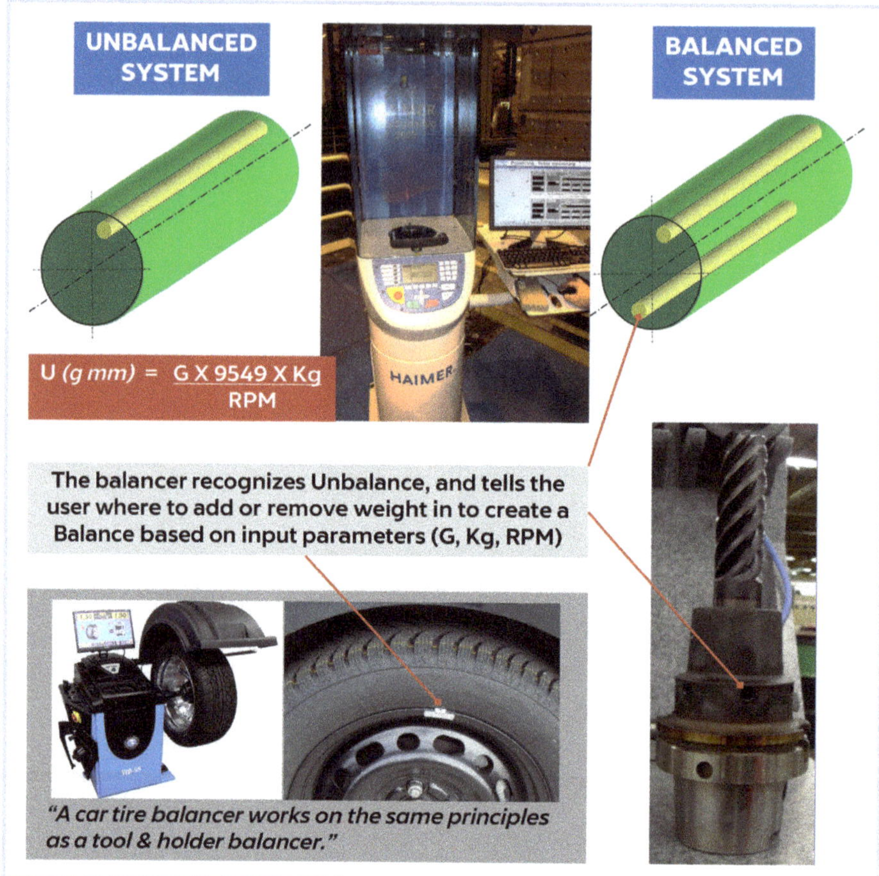

Figure 122. The Balancing Machine – Operating Principles.

16.4 FORCE DUE TO UNBALANCE

Now that we have a firm understanding of Allowable Unbalance U, it is important to realize the consequence of this imbalance can be only felt by the forces it creates when it is spinning. This is labeled as the Force due to Unbalance F_U. ***Figure 123*** shows the complete derivation of the F_U (a centripetal force). Special attention is given to showing numerous conversion factors that are involved in calculating F_U in pounds.

$$F_U = U \times \left[\frac{RPM}{9549}\right]^2 \times \frac{1}{4.45} \qquad [lb]$$

$$F_U(lb) = m.r.\omega^2$$

$$F_U(lb) = U.\omega^2$$

Where ω is the rotational velocity in radians/sec.

$$\omega^2\left[\frac{2\times\pi\times RPM}{60}\right]^2 = \left[\frac{RPM}{9.549}\right]^2$$

Also, since U is in (g.mm), we need to convert it to (kg.m).
This gives us F_U in Newtons.

$$F_U(N) = \frac{U}{1000\times1000}\times\left[\frac{RPM}{9.549}\right]^2$$

$$F_U(N) = U\left[\frac{RPM}{9549}\right]^2$$

Finally, to convert F_U from Newtons to lb, we divide by 4.45.

$$F_U(lb) = U\times\left[\frac{RPM}{9549}\right]^2\frac{1}{4.45}$$

Figure 123. Force due to Unbalance.

16.5 WHEN TO NOT BALANCE - [THE 10% RULE]

When it comes to implementation of balancing the two questions to ask are:

1. What level of precision is needed for the balancing?
2. Can that level of balancing be achieved?
3. Can that level of precision balancing be repeated?

The answer to the first question can be arrived at from the G-Scale, and or from other production relevant archives. If the answer to the second & third question is 'No', then balancing can be an exercise in futility. When the answer to these questions is not so obvious, and the user wishes to make a data-driven decision, the "10% Rule" can be a useful tool.

16.5.1 THE 10% RULE

Recognize that balancing is an output of Acceptable Unbalance U, otherwise understood as "how much unbalance can you live with?" There is always going to be dormant unbalance residing in any balance system. The question to ask is, whether the forces due to that unbalance are going to contribute detrimentally during operation. The 10% rules states that: "Balancing is not necessary if the force F_U created by the Maximum Allowable Unbalance U is within 10% of the Total Cutting Force F_T", ***Figure 124***. Simply put we are evaluating if the cutting force F_T is high enough to where it renders the Force due to Unbalance F_U to be insignificant.

When balancing will not work!

THE 10% RULE:
Balancing is not necessary if the force (F_U) created by the maximum allowable unbalance (U) is within 10% of the total tangential cutting force (F_T).

$$U(g\,mm) = \frac{G \times 9549 \times Kg}{RPM}$$

$$F_U(lb) = U \times \left[\frac{RPM}{9549}\right]^2 \frac{1}{4.45}$$

$$F_T(lb) = \frac{HP \times 33000}{SFM}$$

If

$$F_U \leq 10\% \text{ of } F_T$$

Balancing is not required

Figure 124. When to Not Balance – The 10% Rule.

16.5.1.1 EXAMPLE OF WHEN TO NOT BALANCE

Figure 125 shows a real-life example of the '10% rule' at work to justify not balancing a system. Here the assembly is that of a large boring bar and holder, that is intended to make a precision hole in Aluminum.

- Based on the cutting parameter the Total Cutting Force F_T is calculated to be 65.7 lb.
- Next, the Acceptable Unbalance U is calculated to be 294.3 g mm. Intuitively, this U value is high as compared to other typical cutting operations. This stems from the relatively high weight, and low RPMs. In other words, the system can accommodate a lot of Unbalance before its forces contribute detrimentally.
- Consequently, the Force due to Unbalance F_U is calculated to be 0.726 lbs.
- Finally, upon comparing the F_U and F_T we see that F_U is 1.1% of F_T. Which is significantly less than 10%.

The above analysis conclusively shows that the system does not need to be balanced. In fact, attempting to balance this system would create a scenario that is not repeatable. Each time a well-intentioned user attempts to balance such a system, the balancing machine would yield varied recommendations of adding or removing mass, thus frustrating the user.

> **CAUTION:** Often times machine builders publish a required balancing spec as part of their warranty clause. This is especially the case for machines that cut soft metals at high RPMs, or those that employ purely High Velocity techniques. The customer should preemptively scan the cutter list for that machine, and by using the calculations for 'the 10% rule' identify the tools that may not be good candidates for balancing. This data-driven decision can help the user have an informed communication with the machine builder, and revise the warranty clause to allow use of some non-balanced tools.

EXCEPTION: Example Boring Bar

EXAMPLE: Consider the below identified Boring Bar, intended for use in Aluminum, and weighs 12.33 kg:

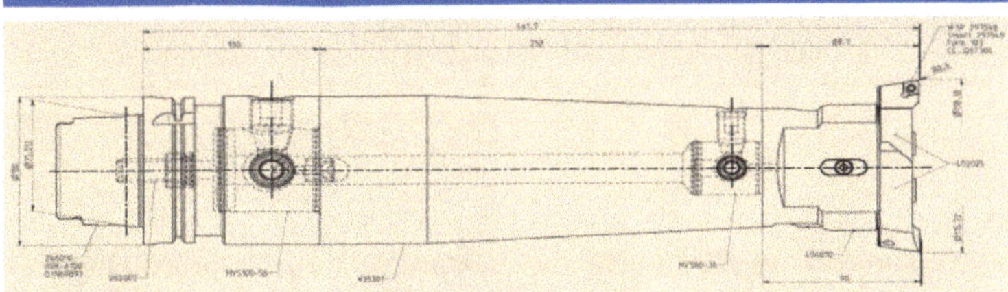

CUTTING FORCE (F_T) CALCULATION
OD = 4.65"
ID = 4.53"
SFM = 1217 ft/min
RPM = 1000 rev/min
IPM = 8.0 in/min
MRR = 6.9 in³/min
F_T = 65.7 lb

MAX. ALLOWABLE UNBALANCE (U) CALCULATION
G = 2.5
Weight = 12.33 kg
RPM = 1000 rev/min
U = 294.3 g mm

FORCE (F_U) DUE TO MAX. ALLOWABLE UNBALANCE CALCULATION
U = 294.3 g mm
RPM = 1000 rev/min
F_U = 3.23 Newtons
F_U = 0.726 lb

$F_U / F_T = 0.726 / 65.7 = 0.011 = 1.1\%$

Thus showing that the force generated by the maximum allowable unbalance (F_U) is 1.1% of the Cutting Force (F_T). Since this 1.1% is well below the established 10% rule, this system does not need to be balanced.

ALSO VISUALIZE

$U\ (g\ mm) = \frac{G \times 9549 \times Kg}{RPM}$	G=2.5	Kg =12.33 [HEAVY]	RPM=1000 [LOW]	U=294.3 [HIGH]

"Therefore, the system can live with or tolerate a lot of Unbalance"

Figure 125. When Not to Balance – An Example of the 10% Rule.

Figure 126, shows other common cutting tool setups that are not balanced. These tools are employed in numerous precision machining operations such as (Top Left to Right): large hole boring, power generation turbines, Crankshaft broaching, tunneling, slotting, precision boring, and a Ferris wheel.

Figure 126. Setups that are Typically Not Balanced.

17

CHATTER & VIBRATIONS ADVANCED DYNAMIC ANALYSIS

In the real-world machine tools are stiff as compared to the tools that go on them, however they are not infinitely stiff. Understanding the dynamic characteristics of the machine and tools becomes especially important when it comes to high velocity machining. This knowledge can then help quantify the vibration, and lead to improvements in cutting performance. An understanding of chatter theories as related to milling cutters shows the scientific means of recognizing a dangerous cutting operation, and optimizing the cutting parameters to eliminate chatter and in many cases enhance the productivity. The conventional thought to slow down the cutting process when chatter is audibly recognized can be detrimental to a cutter, especially when dealing with high velocity machining.

In this chapter, an approach is presented for achieving chatter free milling, and optimizing metal removal rates for industrial metal cutting applications. Techniques including acoustics, computer simulations, and Finite Element Analysis (FEA) are shown for chatter identification, and process optimization. It is revealed that each technique plays a unique role in process optimization of high-speed machining. A methodology is presented for applying these techniques individually and in combination. Actual cases are presented which highlight the information received and performance gains achievable. It is also shown that prior to any actual machining, it is possible to design tools for stability around a machine system, and provide machining optimization offline.

17.1 TYPES OF VIBRATIONS

This section gives an understanding of the common types of vibrations. Chatter is then recognized as a very specific type of vibration. Specific attention is then given to understanding Chatter theory. This basic knowledge is simply intended to form a substantial backbone towards using the available tools to detecting and solving chatter problems in milling.

As shown in ***Figure 127***, three types of vibration are:

- Free Vibration.
- Forced Vibration.
- Regenerative Vibration.

A brief explanation of each of the three is provided in the following subsections.

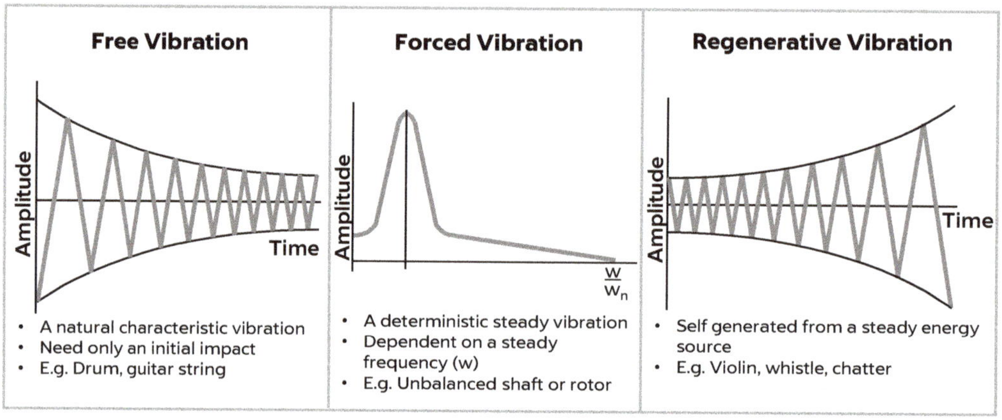

Figure 127. Free, Forced, and Regenerative Vibrations.

17.1.1 FREE VIBRATION

Free vibration is a characteristic natural vibration of a system. Some simple examples of free vibration are a drum, or a guitar string. Free vibrations only need an initial impact to get them started. Also as seen in ***Figure 127***, the amplitude of the vibration decays with time.

17.1.2 FORCED VIBRATION

Forced vibration is also termed as a steady deterministic vibration. An unbalanced shaft or rotor are typical causes that lead to forced vibration in a system. The forced vibration as seen in ***Figure 127*** is termed as a Frequency Response Function (FRF), where ω is the frequency of the exciting force and ω_n is the natural frequency of the system. The natural frequency is represented as:

$$\omega_n = \sqrt{\frac{k}{m}}$$

Where k is the stiffness and m is the mass of the system.

Forced vibration happens at the frequency of the exciting force. The amplitude of motion depends both on the amplitude of the force, and on the frequency of the force. A low frequency of excitation force causes a displacement determined by the familiar "static" stiffness (F=kx). As the excitation frequency increases so does the amplitude of the displacement, up to the "resonance". At resonance the frequency of the excitation force matches the natural frequency. At resonance the amplitude of the displacement is much larger than at low frequency. For excitation frequencies higher than the natural frequency, the amplitude of the displacement decreases.

17.1.3 REGENERATIVE VIBRATION [CHATTER]

Regenerative vibration occurs when a steady input of energy is in some way modulated into vibration. Some examples of regenerative vibration are a whistle, a violin, and chatter. As seen in ***Figure 127*** the amplitude of this type of vibration increases with time. Being that chatter is a regenerative vibration, this concept is dealt with in much greater detail in the subsequent sections.

17.2 WHAT IS CHATTER

In metal cutting, ***Chatter*** is a regenerative vibration that occurs between a tool and a workpiece. Recognize that RPM (revolutions per minute) is directly related to Frequency Hz (cycles per second). The cycles created by one revolution of a cutter is broken up by the number of edges. The rest is just a conversion factor from seconds to minutes. Chatter can be detrimental to surface finish, and tool life. The RPM corresponding to a particular Chatter Frequency can be calculated by the following expression:

$$RPM = \frac{Hz \times 60}{\#\mathrm{T} \times N_{1,2,3,\ldots}}$$

Where:

Hz = Is the Chatter Frequency.

#T = Number of Edges

N = Lobe Number. Ranging from 1, 2, 3, 4 onward. These are basically the Harmonics of the frequency. It should also be noted that the smaller the N, the larger the likely hood of success. (ie larger the pockets of stability). This will be further explained in the section on Stability Lobes. The primary cause of chatter is regenerative waviness, which is explained in the following sub-section.

17.2.1 CHATTER THEORY

Chatter Theory states that "Chatter cannot exist when the Chatter Frequency is equal to the Tooth Passing Frequency or one of its harmonics". The big learning now is: How to detect the Chatter Frequency. This is pretty much what a lot of the mathematical complexity revolves around.

Now that chatter is identified as a very specific type of vibration (i.e. Regenerative), very specific theory needs to be applied to resolving it.

> **CAUTION:** In the world of metal cutting the other types of vibrations (i.e. Free, and Forced) also exist during machining. Apply Chatter Theory to these vibrations will not solve the problem. They have to be dealt with their own unique solutions. The intent in learning Chatter Theory is that the user can differentiate the type of vibration problem they are trying to solve, and apply the appropriate instruments and techniques. **For Example**: If a spindle bearing is worn down unevenly, or a Facemill was not securely tightened to the holder, it would create a type of vibration know as Forced vibration. These Forced Vibrations could not be solved by using the Chatter Theory, but the Sensor-Based system will be able to identify them.

17.3 THREE WAYS TO DETECT & SOLVE FOR CHATTER

As seen in ***Figure 128***, are three industry standard techniques to analyze chatter, and solve for eliminating it.

- Acoustic Based Chatter Detection System.
- Sensor-Based Dynamic Testing (Hammer or Tap Testing).
- Finite Element Analysis (Advanced Dynamics / Computer based simulation).

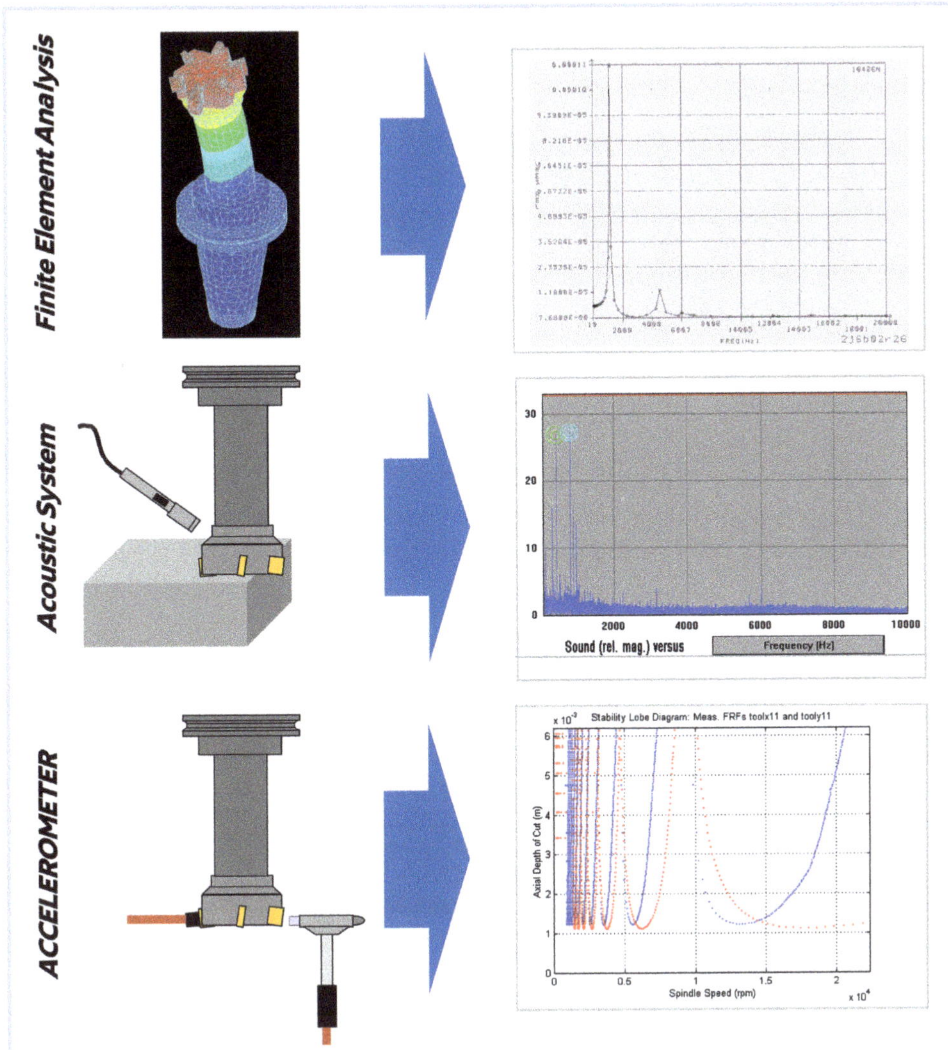

Figure 128. Three Systems to Solve for Chatter Problems.

Each of these above techniques come with their specific instrumentation, as well as degrees of difficult in usage. To start with, as seen in the Flowchart in ***Figure 129***, there are some basic questions that could help the user determine which technique to select:

Q1) Is the System Built?

If the answer is NO, then the only option is to use a computer-based system with Finite Element Analysis. As one can imagine, such a system is only as good as level of maturity of the simulation software, and the correctness of the system modeling.

Q2) If the System Is Built, Then Is It Cutting?

I the answer is NO, then we can simply assemble the tool in the holder, and install it in the machine spindle. This is where the user can us Sensor-Based Tap Testing technique. This is also the most demanding approach in terms of background understanding of Advanced Dynamic theory. However, modern day software & instruments such as make it quite user friendly.

Q3) Is the System Cutting in the Machine?

If the answer is Yes, then the user can implement the audio-based system (such as the the acoustic based chatter detection system. This is also the easiest and most user-friendly system. However, this system is strictly focused on listening for a chatter frequency, and making a singular recommendation of RPM to eliminate it. Audio based analysis can also be very limited in its diversity of vibration data outputs.

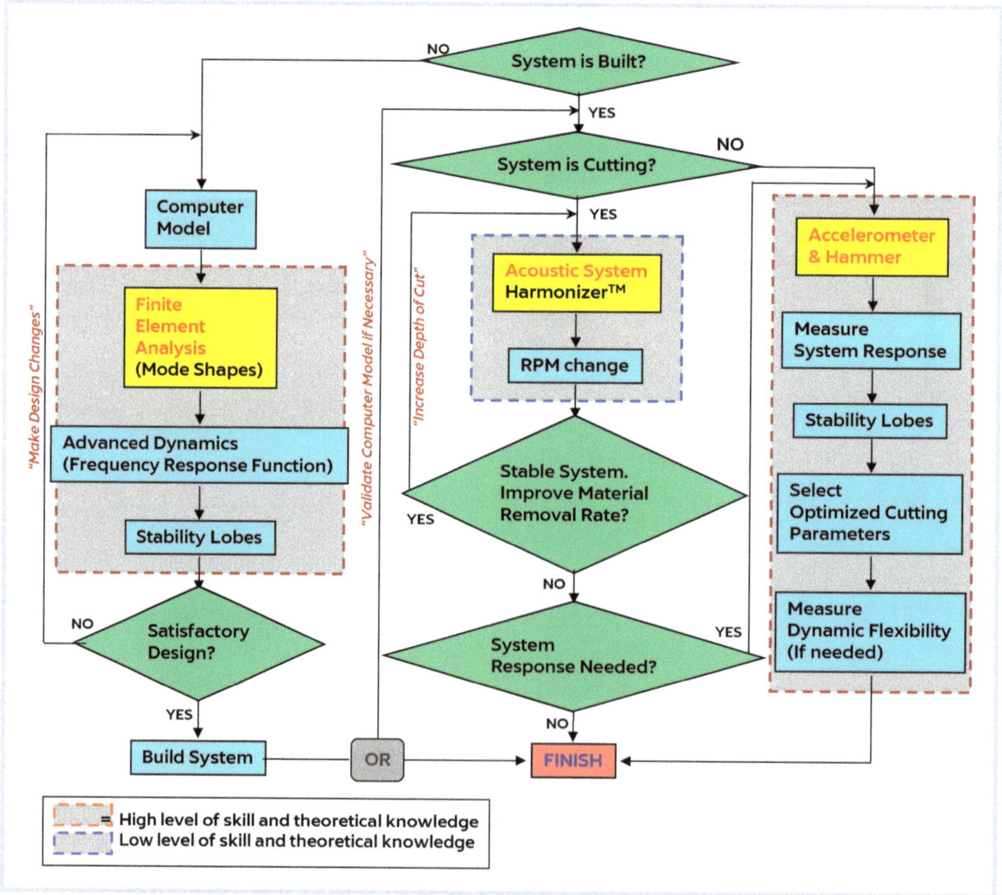

Figure 129. Selecting the Most Optimal Chatter Detection System.

In the subsequent section we will discuss in detail the components of the Advanced Dynamic Theory applied. We will also subject each of these techniques to an **Example Tool**. This will allow the user to track the pros, cons, and degree of difficulty in the use of each of these techniques.

17.3.1 ACOUSTIC BASED CHATTER DETECTION

The Acoustic Based Chatter Detection is a PC based acoustic chatter detection system that with the aid of a unidirectional microphone is capable of detecting chatter, and then makes RPM recommendations so as to eliminate the chatter, ***Figure 130***. The RPM at which a cutter spins, combined with the cutter diameter and number of teeth is responsible for creating the tooth passing frequency. Based on the engagement conditions, the relative magnitude of sound is recorded, and the level of cutting stability is recognized. The Acoustic Based System follows a strict set of rules so as to eliminate the regenerative waviness and hence eliminates the existing chatter by making an RPM recommendation which would put the system in a stable cutting region.

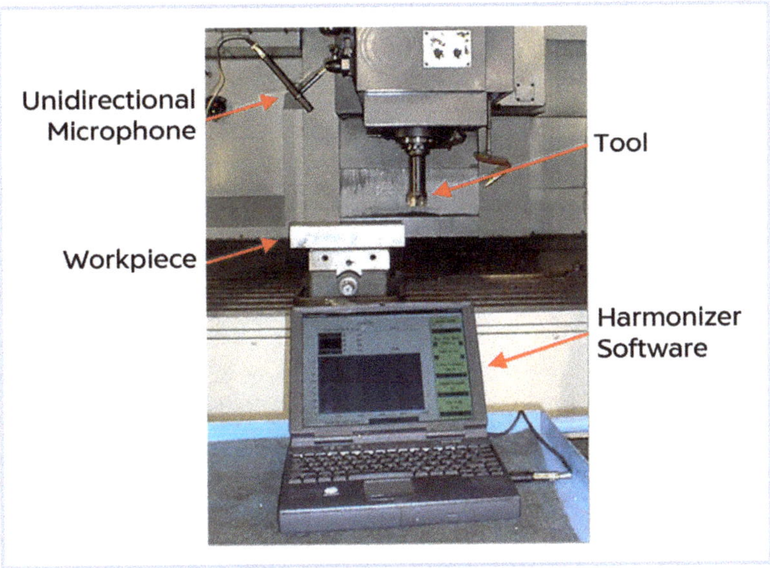

Figure 130. Setup for the Acoustic Chatter Detection System.

17.3.1.1 PROCEDURE FOR ACOUSTIC BASED CHATTER DETECTION

The Acoustic Based Chatter solution is perhaps the least theory intensive. It is a very user-friendly system, that is designed to provide a focused solution for chatter specifically. Below is a high-level breakdown on how it works:

- **USER INPUT:** The user has to tell the system the cutter dimensional detail, and its operating current parameters (DIA, #T, RPM, IPM, Workpiece Material, etc). The user also has the ability to enter boundary conditions (Max RPM of Machine, Max Speed for that Carbide, etc). In doing so, the system will make a recommendation within the boundary conditions.

> **NOTE:** Based on the above user input the system already knows the Once Per Revolution Frequency, Tooth Pass Frequency and its harmonics. It does so by the following equation:
>
> $$Hz = \frac{RPM \times \#T \times N_{1,2,3,\ldots}}{60}$$

When #T=1, it gives us the Once Per Revolution Frequency. When #T is the actual number of cutter edges, it gives us the Tooth Passing Frequency. As N values are used from 1, 2, 3, onward; it gives us the first, second, and third Harmonics respectively.

- **RECORD THE SOUND:** Start the cut at the machine, and once the Cutting Tool is completely in the cut, start recording the sound with the unidirectional microphone. Hopefully, there is chatter in the cut.
- **FILTERS:** It should be recognized that the Acoustic Based Systems have patented advanced audio filters. It is these filters that are able to separate out the Chatter Frequency from all the other frequencies in the cut (including the ones calculated earlier: Tooth pass frequency, and once per revolution frequency). This unique ability to identify the Chatter Frequency, is what makes the guts of the differentiation of the Acoustic based system.
- **IMPLEMENT CHATTER THEORY:** Now that the Acoustic based system has identified the Chatter Frequency, it makes an RPM recommendation based on Chatter Theory. This theory states that "Chatter cannot occur when the Chatter Frequency is equal to the Tooth Pass Frequency or one of its harmonics". This RPM recommendation is calculated as:

$$RPM = \frac{Hz \times 60}{\#T \times N_{1,2,3,\ldots}}$$

Where, Hz is the Chatter Frequency detected by the Acoustic system.

> **NOTE:** As the N values change from 1, 2, 3, onward; the RPMs calculated for each decrease in values.

- **MAKES RECOMMENDATION:** Now that the Acoustic System has all the suggestions of RPMs stored in its memory, it then compares them with the previously established boundary conditions. The system then finally makes a recommendation of the highest RPM that fall within the boundary conditions.
- **ADJUST CUTTING PARAMETERS:** The user would then incorporate the newly recommended RPM, and accordingly recalculate the Feedrate (IPM) so as to keep the original Chipload (APT).

$$IPM = APT \times \#T \times RPM$$

Finally make the new cut, and re-record the sound. Hopefully, the chatter is eliminated. If not, continue with the new recommendation and allow the system a few tries to dial in the most stable cutting region of RPM. If the system just cannot stabilize the Chatter (Regenerative Vibration), then you are probably dealing with a Forced Vibration, and additional Advanced Dynamic Analysis (Tap Testing) is required.

The Example Tool: *The sample tool subjected to analysis by each of these three techniques is shown in* ***Figure 131****. This is a* **2.5" diameter Facemill***, with* **6 effective edges***, making a channel cut (***WOC=2.5"***) in workpiece of material* **6061 Aluminum.**

As seen in ***Figure 131***, the cutter in our investigation chattered at 1656 RPM. It detected a Chatter Frequency of 449Hz. Based on the Acoustic Based System's recommendation this cut was stabilized at 4490 RPM. In order to keep the chip load constant, the feed rate is increased from 50 in/min to 135 in/min. Not only was the Acoustic Based System able to stabilize the cut, but it also yielded an improvement of 170% in the material removal rate. The Acoustic Based System makes its recommendations based on the principle explained in ***Figure 131***. Simply put the Acoustic Based System tries to match the chatter frequency detected to a tooth passing frequency or a multiple of the tooth passing frequency. It accomplishes this by making RPM recommendations that in turn changes the tooth passing frequency, so as to match the detected chatter frequency.

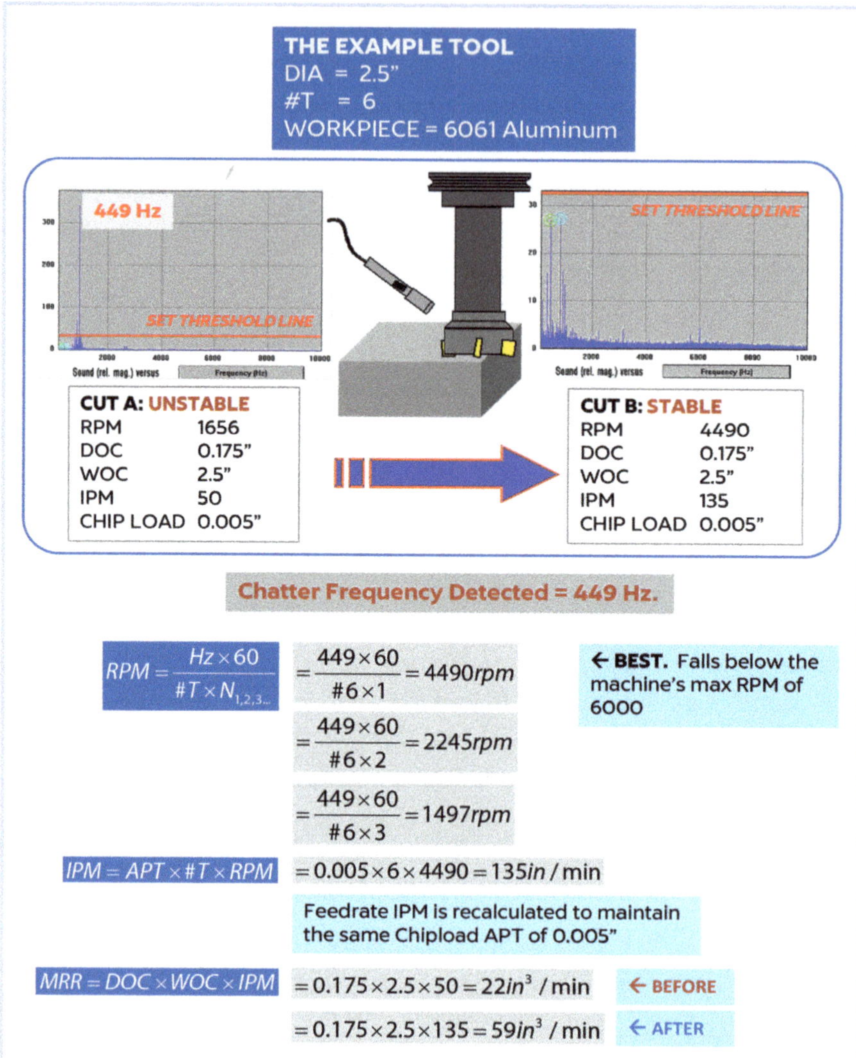

Figure 131. Using the Acoustic Based System to Stabilize a Cut.

17.3.2 SENSOR-BASED CHATTER DETECTION SYSTEM (HAMMER OR TAP TESTING)

Before we go any further with the complex mathematics of Chatter Theory, it is best to get familiar with the Stability Lobe Diagram, ***Figure 132***. In fact, the Stability Lobe Diagram is the end result or output of applying the Chatter Theory to any cutting tool & holder assembly.

The Stability Lobe Diagram is the user-friendly output of the vibrational data collected on a cutting tool, holder and machine spindle assembly. The real-life vibrational data is subjected to complex mathematical manipulation to yield a graph that is relatable to a machinist relevant parameter. As shown in ***Figure 132***, the Stability Lobe Diagram is a plot of the RPM vs. DOC. The regions shown inside the lobes (blue) are considered Unstable. Chatter will always occur in these Unstable regions. Pockets of Stability are identified in between these Unstable Lobes. Chatter will not occur in the Stable pockets. Additionally, an imaginary line can be drawn at the bottom vertex of all the lobes. This is known as the b_{Limit} line (or Limit DOC line). It should be noted that the system is always Stable (Chatter Free) below the b_{Limit} line.

CAUTION: The Stability Lobe Diagram is an output plot of RPM vs DOC, while all other machining parameters are held constant (INPUTS). These parameters include: WOC, APT, and workpiece material. Changing these input parameters would yield a different Stability Lobe Diagram.

NOTE: The subsequent sub-section deep dives the mathematics of vibration theory needed to create the Stability Lobe Diagram. These are mostly shown in this text for the sake of completeness. In modern day application, a user need not learn the development of this Chatter Theory. They can simply develop Stability Lobe diagrams with the use of Sensor-Based equipment, and related software. However, it is strongly recommended that the user be very familiar, and is confidently able to clearly interpret the Stability Lobe Diagram itself.

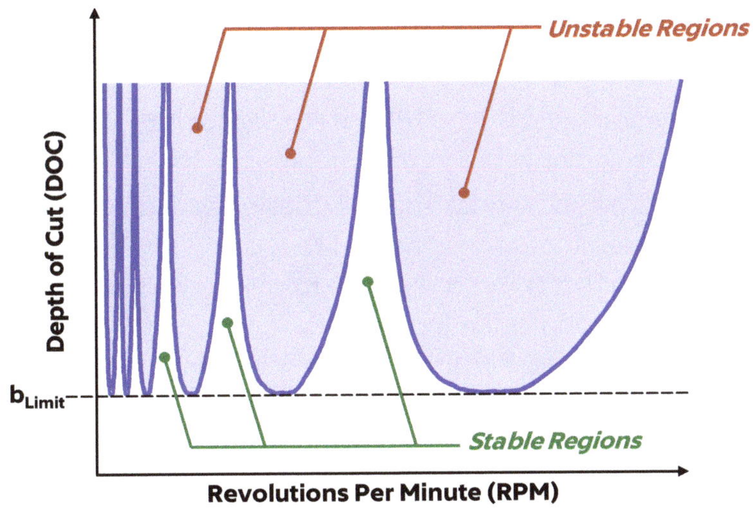

Figure 132. Stability Lobe Diagram.

17.3.2.1 ADVANCED DYNAMIC ANALYSIS TO DEVELOP STABILITY LOBE DIAGRAMS

Now that we realize that the Stability Lobe Diagram is the end-game in our Sensor-Based analysis of Chatter, let's focus our attention on the step-by-step process it takes to generate it.

17.3.2.1.1 Procedure for Sensor-Based Stability Mapping

Figure 133, show the setup for conducting the Sensor-Based Advanced Dynamic Analysis. It should be noted that this Sensor-Based system not only solves for chatter, but maps out the entire dynamic characteristics of the cutting tool & holder assembly by way of Stability Lobe Diagrams:

- **TOOL & HOLDER SETUP:** The cutting tool is assembled to the holder, and then needs to be mounted in the machine spindle. The machine spindle should be set to a neutral condition to where it is freely revolving.
- **SENSOR & INSTRUMENTATION SETUP:** The accelerometer (which is a piezo-electric sensor) is mounted at the tip of the cutting tool as shown in ***Figure 133***. It is positioned at the tip of the cutting tool so as to collect dynamic data at the location where cutting occurs.

The Hammer as shown in ***Figure 133*** is also a sensor type setup. It is made ready so as to strike the tip of the cutting tool directly across from the Sensor location (ie 180° apart).

- **TAP OR STRIKING THE TOOL:** When the hammer with its internally calibrated sensor strikes the cutting tool, this creates the known INPUT. The Sensor sitting directly across on the other side of the cutting tool is then picking up the OUTPUT provided by the Hammer's INPUT. Recognize that in between the INPUT of the Hammer, and the OUTPUT of the Sensor, is basically the vibration tendencies of the cutting tool & holder mounted to the spindle.
- **DATA COLLECTION:** By way of specially designed electrical connections, the Sensor and Hammer are connected to a computer laptop. The laptop comes equipped with advanced software to collect this vibrational data. The user is required to enter the cutting tool description, cutting parameters, and workpiece material details into the software. Additionally, the user also has to inform the software on the relative position of the sensor & hammer strikes (Direct vs Transverse).

Once the data comes into the software, it is first presented to the user as a Frequency Response Function (FRF). The next several sections will walk the user at a high level through converting an FRF to a usable Stability Lobe Diagram. Of course, the software does so with just a few clicks of the mouse. However, for the sake of completeness a high-level explanation is provided.

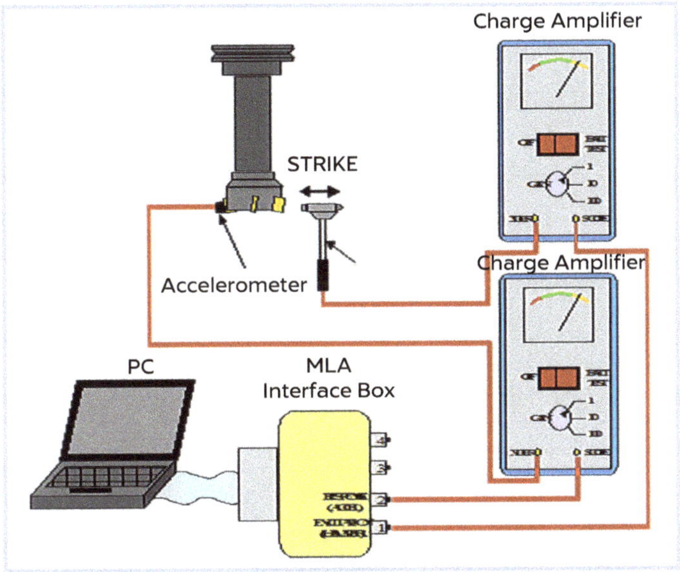

Figure 133. Sensor-Based System for Chatter Analysis.

17.3.2.1.2 Frequency Response Function (FRF) & Mode Shapes

The ***Frequency Response Function (FRF)*** describes how a system will vibrate in response to different frequencies of excitation, ***Figure 134***. The FRF is a measurable function, and can be used to compare and predict the performance of cutters, and machine tools. There is a very high correlation between the FRF and the amount of speed and power that can be used in a milling operation. Whereas single degree of freedom systems has one natural frequency, multiple degree of freedom systems have one natural frequency for each degree of freedom. Each natural frequency has a corresponding characteristic deformation pattern (**Mode Shape**), ***Figure 134***. Vibration in multiple degree of freedom systems may be thought of as a sum of vibrations in the individual modes.

> **NOTE:** Usually (but not always) the first few **Mode Shapes** (Mode 1, 2, or 3) are the most critical bending modes. Their frequencies are typically the ones that contribute the most detrimentally towards vibration problems. Their amplitudes also tend to stand out. The analysis shown in the subsequent sections will have this priority as well.

Typically, there are two common industry standard techniques to creating FRFs:

- Sensor-Based system such as Tap Testing.
- Computer based simulation such as Finite Element Analysis (FEA).

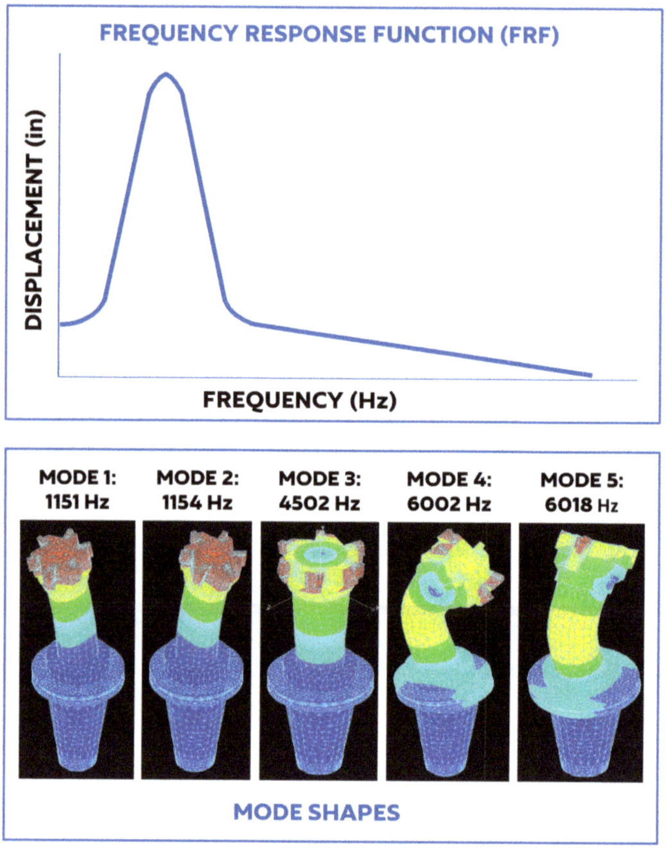

Figure 134. Frequency Response Function (FRF) and Mode Shapes.

17.3.2.1.3 Regenerative Waviness in a Chip

The most undesirable vibrations in milling are regenerative chatter vibrations. The mechanism that causes a steady input of energy from a spindle drive into vibration at the cutting edge is called ***"Regeneration of Waviness".*** When you look at the texture of chip under a microscope, it shows some inherent waviness, ***Figure 135***. When a milling cutter makes a pass, it leaves a specific wavy surface on the work-piece. When the subsequent pass is made, the cutter is now removing material from a previously existing wavy surface, and at the same time leaving behind a new wavy surface. The chip that is created by this cut carries both the waviness from the previous pass, and that translated over by the current pass, ***Figure 135***. If the new cut leads to a chip with variable chip thickness (i.e. waves are out of phase), this would translate as variable forces on the cutting edge, and eventually as vibrations. If the waviness of the chip is in phase it creates a stable cut. In other words, if the chip is in-phase, we are creating the chip thickness as intended dimensionally.

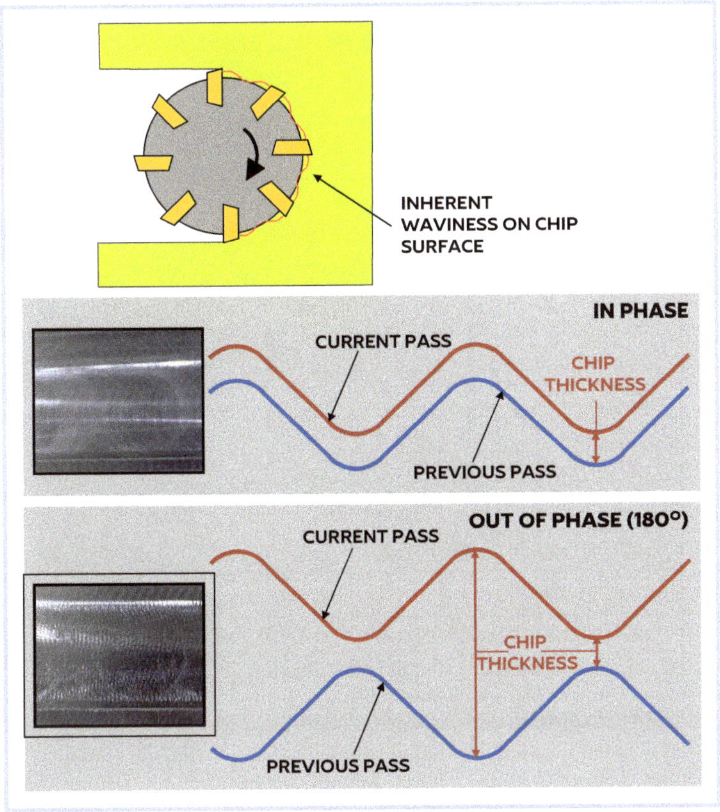

Figure 135. Regenerative Waviness in a Chip.

NOTE: It is important to recognize that the Regenerative Waviness is the dominant cause of chatter, and is not the same as resonance. In most cases Regenerative Waviness is excited well before the resonance of the system gets excited.

Since in these sections we are talking about a Sensor-Based vibration mapping technique, where the cutter has not yet made a cut; clearly, we would not know the Chatter Frequency yet. However, since our goal is to create a vibration characteristic map (Stability Lobe Diagram), we will incorporate Regenerative Waviness (number of waves per chip = 1, 2, 3, 4....) into the computation of the Lobe Numbers (N = 1, 2, 3, 4....).

17.3.2.1.4 Real and Imaginary Parts of FRF

A frequency response function (FRF) can be further evaluated as real and imaginary parts as shown in ***Figure 136***. The equations that led to the generation of the real and imaginary graphs are also identified in ***Figure 136***. Each of these plays a significant role in the development and understanding of chatter theory. The real part of the FRF helps calculate Damping, and the imaginary part helps to calculate stiffness. The section on Stability Lobes helps explain in greater detail the necessity, and nature of the real and imaginary parts of the FRF.

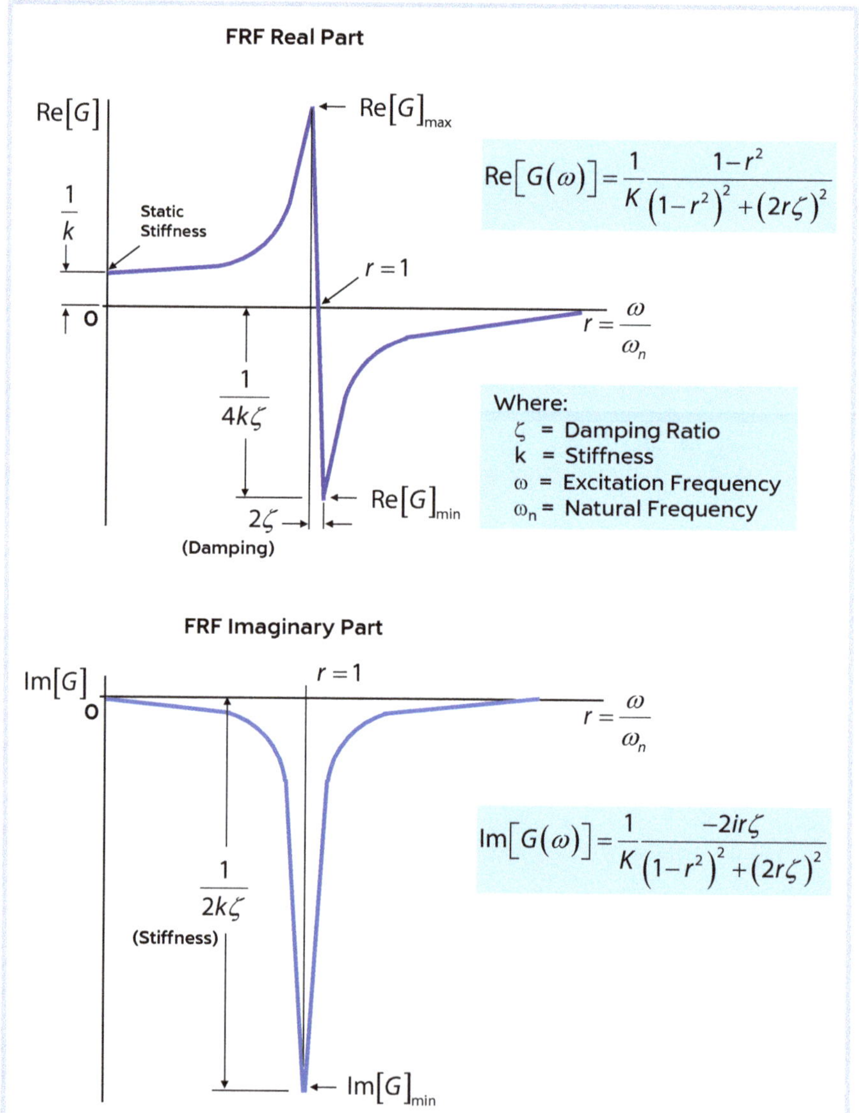

Figure 136. Real & Imaginary Plots of the FRF.

These Real & Imaginary plots can be quite intimidating to create mathematically. Don't worry, the software will create these automatically from the FRF. The intent in this text is to show that the mathematical pathway for an FRF to a Stability Lobe Diagram, is through the Real & Imaginary plots.

17.3.2.1.5 Creating Stability Lobes

The negative part of the Real plot of FRF (***Figure 137***), combined with the number of waves (N) in between subsequent teeth and their phase, leads to the creation of ***Stability Lobes, Figure 138.*** The negative region of the real part of the FRF is used to generate a single lobe. Every frequency point on the negative part of the Real FRF corresponds to a point on the Stability Lobe, as shown by the dotted lines.

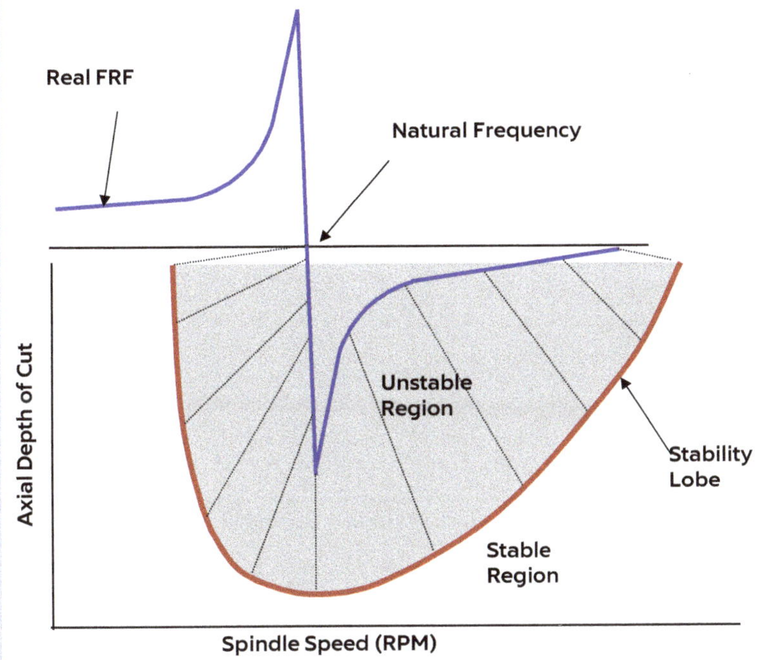

$$b_{lim,crit} = \frac{-1}{2K_s \mu \mathrm{Re}[G]_{min}}$$

Figure 137. Formation of a Stability Lobe from the Real Part of the FRF.

Stability Lobes permits the selection of dramatically larger axial depths of cut at high spindle speeds (***Figure 138***). Large stable gaps occur at spindle speeds where the rotational frequency of the tool or workpiece is equal to the dominant natural frequency of the system. Recognizing the limitations of the system

(Machine & Carbide) and selecting a stable region to work in, can greatly enhance a metal cutting process. Also, the largest stable gaps occur where the tooth passing frequency is equal to the natural frequency of the system. It is at this time that the surface waves and the tooth vibration is in phase. As seen in ***Figure 138***, as the number of waves between consecutive teeth (N) increases, the area of the stable region decreases.

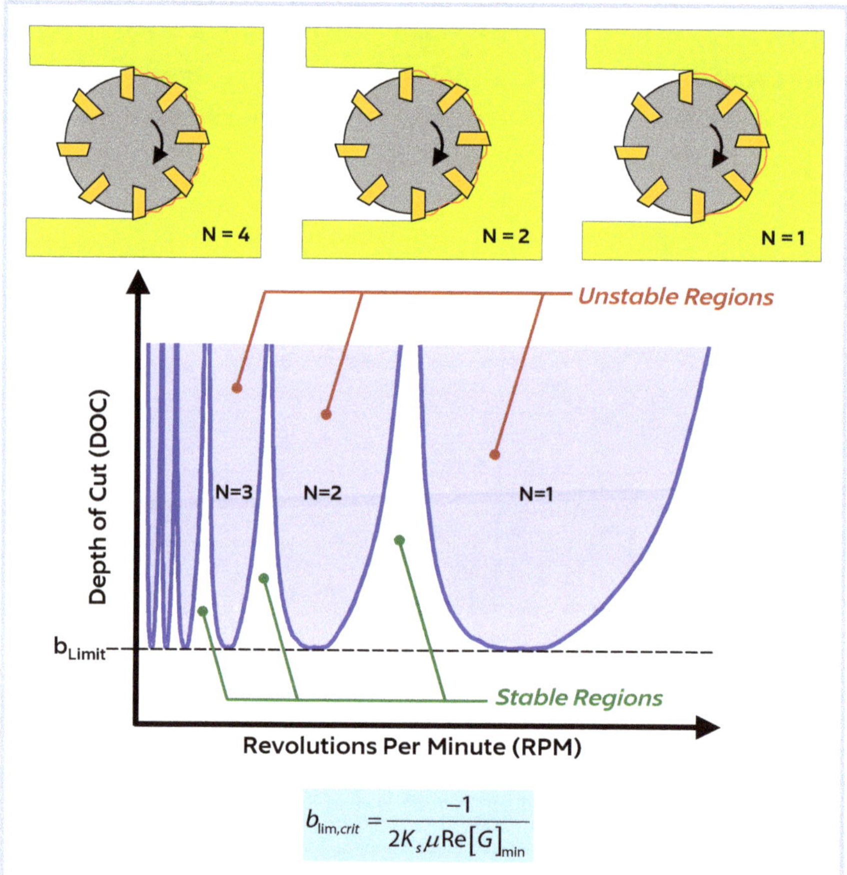

Figure 138. Number of Waves (N) Generated Between Teeth & Stability Lobes.

Upon closer observation of the Stability Lobes, it is evident that maintaining an axial depth of cut below the b_{limit} will always yield a stable cutting region. The negative peak from the Real part of the FRF is used to calculate b_{limit} by using the expression:

$$b_{lim,crit} = \frac{-1}{2K_s \mu Re[G]_{min}}$$

Where:

K_s = cutting stiffness

μ = Directional Orientation Factor (μ = 0.34, for slotting)

$Re[G]_{min}$ = Negative peak of the Real part of FRF.

As seen in ***Figure 139***, μ is the directional orientation factor. In a slotting operation the cutting force usually lags by 20°. Thus μ for a slotting operation works out to 0.34.

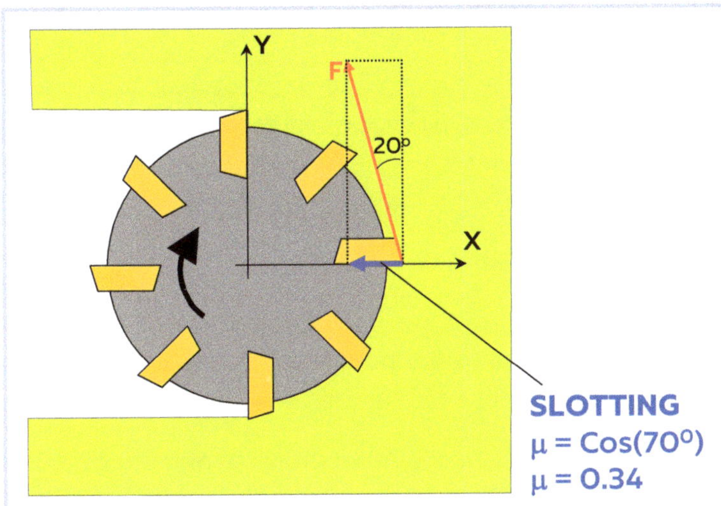

Figure 139. Directional Orientation Factor μ for Slotting.

It is also seen that these regions extend up to greater depths of cuts in between the shaded unstable pockets. The region between N=1 and N=2 shows a greater area, and in turn greater variability of RPMs (in the stable regions) than that between N=2 and N=3. It may not always be possible to take advantage of the largest stable region. This could be limited by the machine's maximum RPM capability, and or by the maximum speed (Surface feet per minute, SFM) that the cutter material can handle.

17.3.2.1.6 The Convergence Technique for Finding Stable Regions

The top plot of ***Figure 140***, shows a stability lobe diagram (hatch area indicates chatter) for a 2-tooth cutter. The bottom plot displays (BLUE bold non-linear lines) the chatter frequency (Hz) as a function of spindle speed. Also on the plot are line of tooth passing frequency (Hz) multiples (RED straight lines marked $2f_t$, $3f_t$,...) as a function of spindle speed.

Chatter Theory states that "Chatter cannot occur when the Chatter Frequency is equal to the Tooth Passing Frequency or one of its Harmonics".

The indisputable concept is that "Spindle speed and chatter frequency are related". Select a spindle speed so that the tooth passing frequency or some multiple of the tooth passing frequency is equal to the chatter frequency.

Referring to ***Figure 140***, assume that a cut is made at point "a" at a 0.5 mm depth. The upper stability plot indicates that this cut is chattering. The chatter frequency can be determined in the lower plot by finding the frequency for the same spindle speed point on the lowest chatter curve (BLUE bold non-linear line). It is about 1650 Hz. Graphically the convergence technique directs the spindle speed change to the spindle speed curve to the two times multiple of the tooth passing frequency ($2f_t$) which corresponds to point "c" with spindle speed of just under 24,750 rpm.

$$RPM = \frac{Hz \times 60}{\# Teeth \times N} = \frac{1650 \times 60}{2 \times 2} = 24{,}750\ rpm.$$

Returning to the upper plot indicates the new speed is in a chatter free area at point "c". Starting at another spindle speed, point "b", results in a similar outcome demonstrating the convergence technique.

On the other hand, should the user start at point "b" and chooses to stabilize the cut at a lower RPM, then they would use N=3 in the calculation:

$$RPM = \frac{Hz \times 60}{\# Teeth \times N} = \frac{1650 \times 60}{2 \times 3} = 16{,}500\ rpm.$$

For N=3, the calculated RPM is 16,500 which corresponds to the next stable pocket down, and is located at point "d".

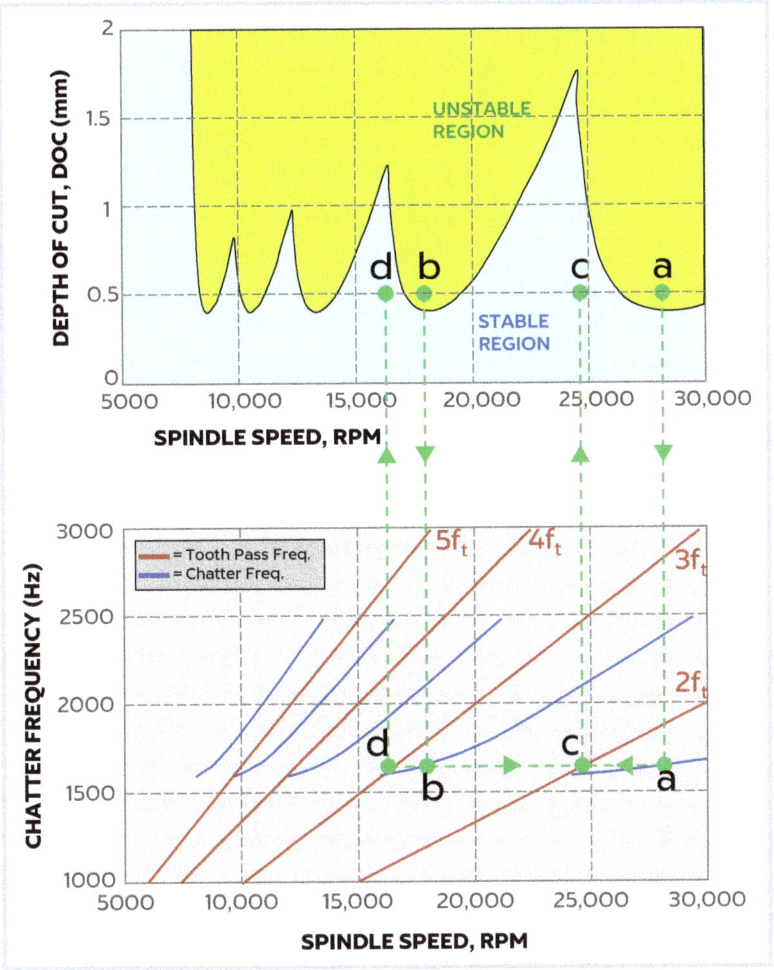

Figure 140. Demonstration of the Convergence Technique for Finding Stable Regions.

The Example Tool: *The same Example Tool that was subjected earlier to the Acoustic-Based system, is now subjected to the Sensor-Based Tap Testing Analysis. The results yield the Stability Lobes as seen in* ***Figure 141****. This particular machine has a maximum RPM of 6000. Special focus should be given to the pockets of stability. In our earlier investigation with the Acoustic Based System a recommendation to of 4490 RPM was made. This recommendation also falls in the first closest stable pocket to the left of the machine's maximum RPM line. Thus, validating the output from the Acoustic Based System. Although the largest pocket of stability exists to the right of the maximum RPM line, this is a limitation of the machine's maximum spindle speed.*

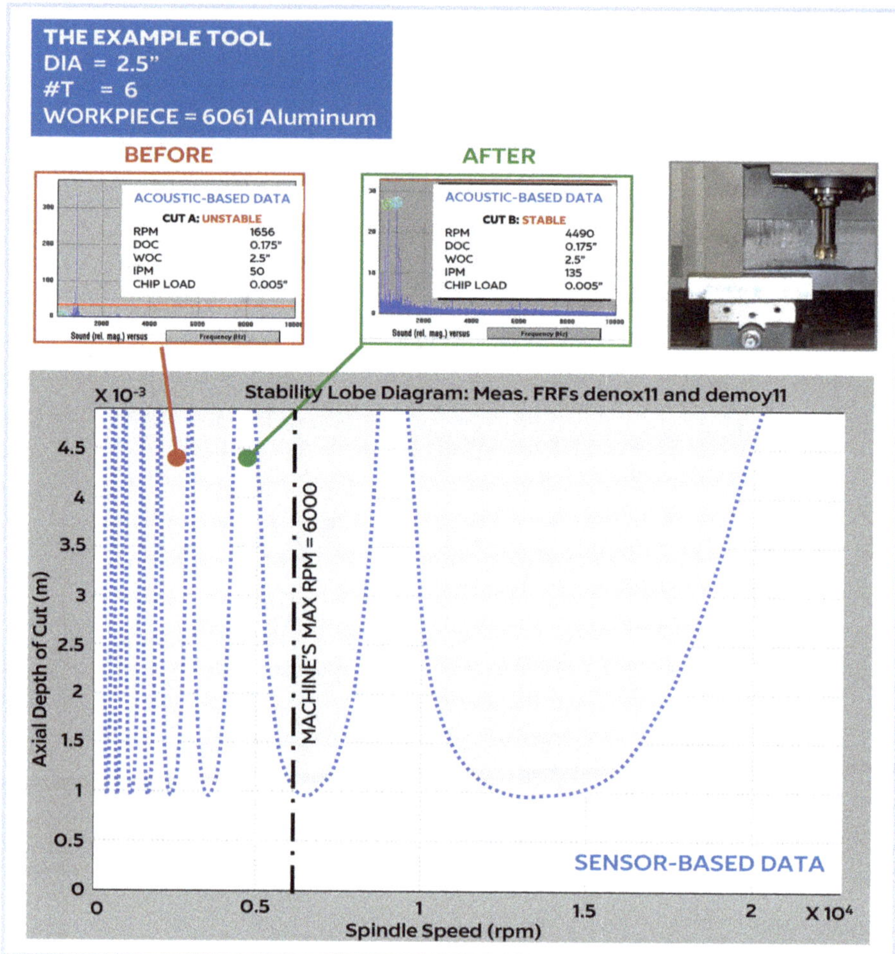

Figure 141. Stability Lobes for the Example Tool from Sensor-Based System, Compared to Results from Acoustic-Based System.

17.3.3 FINITE ELEMENT ANALYSIS (ADVANCED DYNAMICS / COMPUTER BASED SIMULATION)

Finite Element Analysis FEA enables the use of computer simulations to solve for chatter problems before the tool is built, ***Figure 142***. It should be recognized that such an analysis is performed off-line, (i.e. prior to building a prototype tool). This allows for intelligent cutting tool design, and an in depth understanding of the tool vibration characteristics. Quantified scientific data allows the end user to make the right choice with not only selecting the right machine, but also with refining the existing cutting parameters, so as to optimize the milling process.

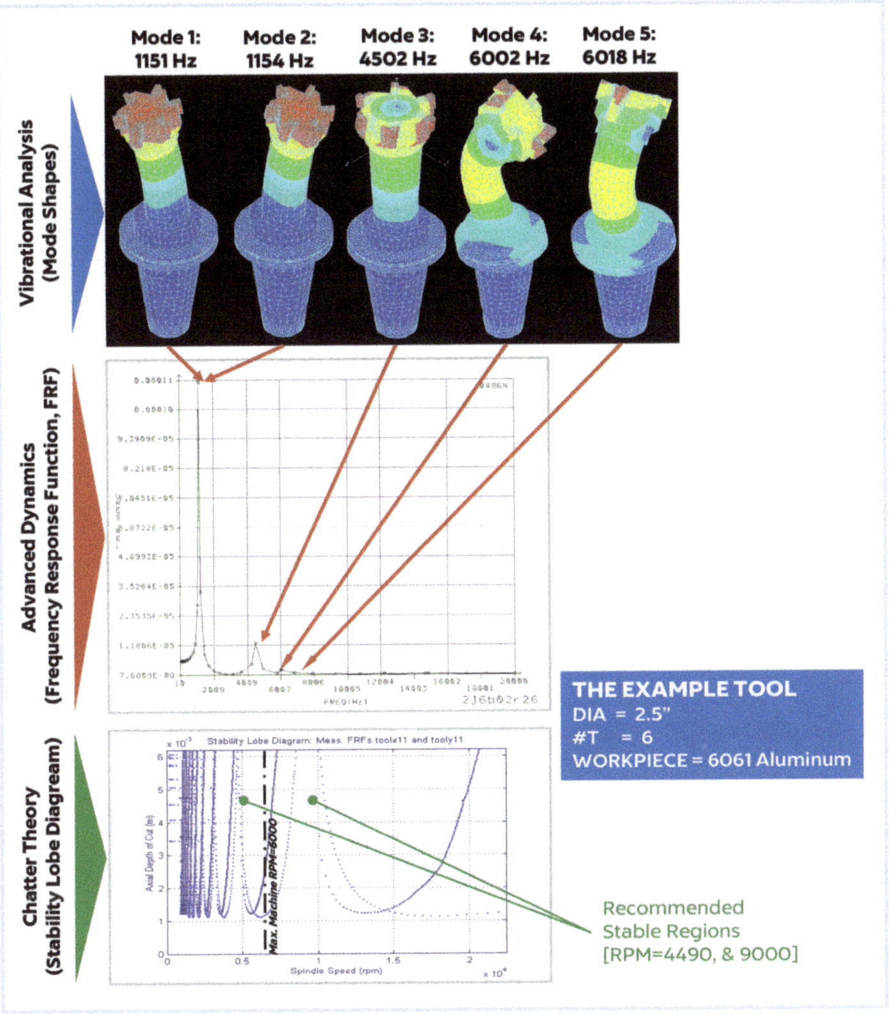

Figure 142. Output from Finite Element Analysis – Advanced Dynamics.

17.3.3.1 PROCEDURE FOR FINITE ELEMENT ANALYSIS IN VIBRATION MAPPING

A basic frequency analysis generates the mode shapes that the system is going through. Next the model is subjected to loading conditions such as cutting force, and its frequency based on the engagement conditions. The output from advanced dynamic analysis is typically seen as a ***Frequency Response Function (FRF)***. This FRF is then subjected to additional mathematical computations to yield ***Stability Lobes***, ***Figure 142***.

Such an analysis is often limited by the computer system that is being used, and requires a high degree of skill to create models that are calibrated to reality. Even though this work may be relatively time consuming from a design standpoint, the time savings during the run-off stages, combined with the confidence level in the product make it well worth the upfront additional time investment.

> **The Example Tool:** *The same Example Tool that was subjected earlier to the Acoustic-Based System and Sensor-Based System, is now subjected to the Finite Element Analysis FEA (**Figure 142**). Most FEA software do not have the capability of creating Stability Lobe Diagrams. Here the FRF output of an FEA system is further subjected to intense manipulation in software that specialize in math computations (like MATLAB or Excel), to create the Stability Lobe Diagram (**Figure 142**, bottom). Pockets of stability appear to be aligned with the results seen in the Acoustic-Based, and Sensor-Based systems. Specifically, RPM=4490 (as recommended by Acoustic System) is the first stable pocket seen to the left of the machine's max RPM line of 6000. If the machine were capable, a larger stable pocket exists at RPM=9000.*

17.3.4 CHATTER FREQUENCY FROM VISUAL INSPECTION OF A WORKPIECE (NO VIBRATION ANALYSIS INSTRUMENTATION)

An unstable cut always leaves a specific finish signature on the workpiece. Simple fact is that, cutting needs to occur in order to recognize, capture or measure Chatter Frequency. This technique does not require any vibration analysis equipment at all. This could be a quick "in the field" type analysis that could help resolve a chatter issue, especially if no vibration analysis equipment were available. It is possible to determine chatter frequency by measuring the distance L in between the chatter marks as seen in ***Figure 143***. This can be done simply with a pair of dial calipers. Once the spacing length L is measured, the frequency of vibration f may be calculated as:

$$f = \frac{V}{5L}$$

Where

f = Vibration Frequency (Hz)

V = Cutting Speed (ft/min)

L = Chatter mark spacing (in)

> **NOTE:** Chatter cannot occur when the Chatter Frequency is equal to the tooth passing frequency or any of its harmonics.

Therefore, once the Chatter Frequency is measured the RPM to stabilize the cut can be calculated as:

$$RPM = \frac{Hz \times 60}{\#T \times N_{1,2,3,4.....}}$$

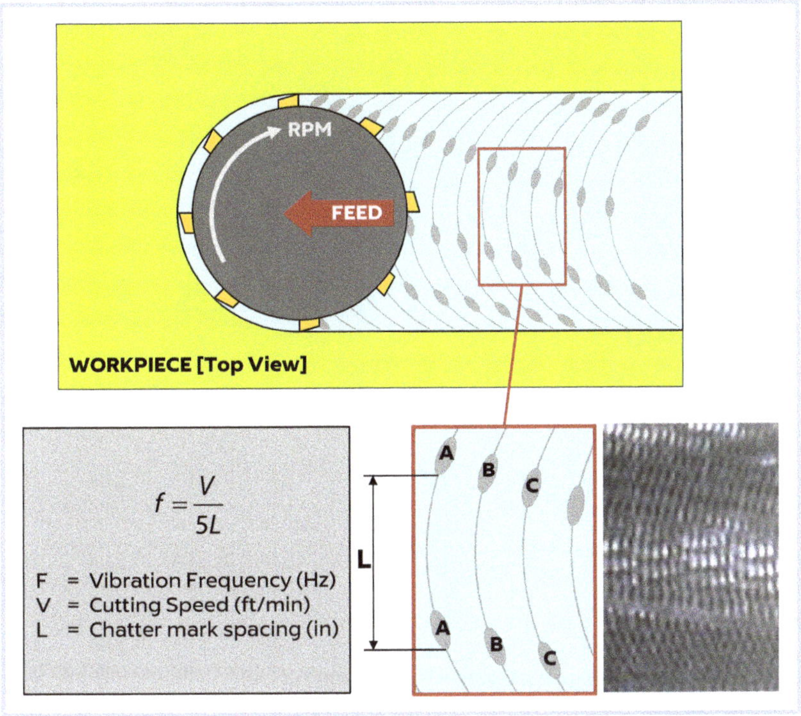

Figure 143. Interpretation of Chatter Marks in Milling.

17.4 ADDITIONAL GUIDELINES FOR RESOLVING VIBRATION ISSUES IN MACHINING

As is obvious to experienced machinists, the elimination of chatter is not strictly dependent upon the use of chatter detection instruments. Issues such as cutter design, tool-path generation, and process optimization, play a vital role in conjunction with chatter detection techniques to optimize the cutting process. Chatter detection tools are mere instruments that can guide the designer, the manufacturing engineer, and the operator to get the most out of a cutting process. Examples of these guidelines is explained in the subsequent sub-sections.

17.4.1 DYNAMIC FLEXIBILITY

For the tool under investigation the system response was measured at six different locations along the body of the cutter as seen in ***Figure 144***. These measurements are taken by the Sensor-Based System, and values are taken from the Real part of the FRF. Recordings were made by striking the hammer in the direction of the machines X and Y axis. The dynamic flexibility at each of the six points for the X and Y direction are generated, ***Figure 144***.

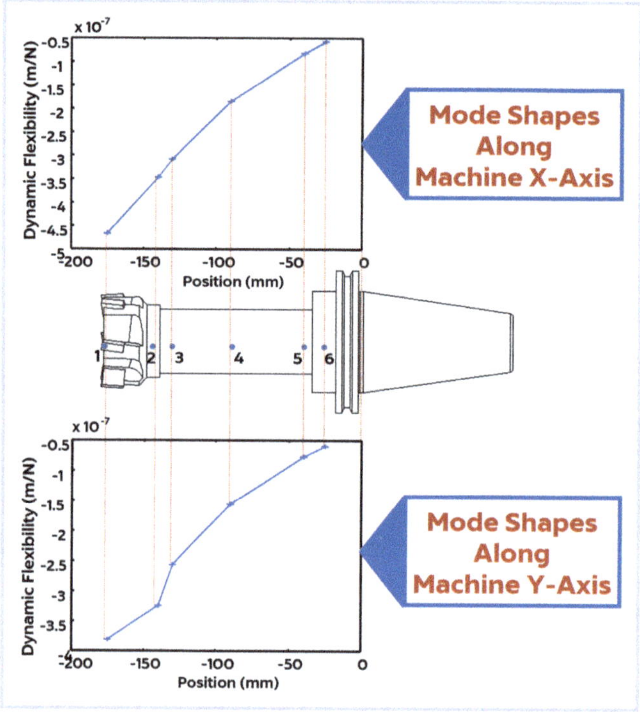

Figure 144. Dynamic Flexibility.

As expected, the flexibility of the system at the gage-line (i.e. Point 6) is the smallest, and that at the tool tip (i.e. Point 1) is the largest. The plot of dynamic flexibility along the Y-axis shows an abrupt change in slope between points 2 and 3, as compared to that of the X-axis which is a smooth curve. This observed abrupt change in slope along the Y-axis occurred because the hammer strikes made at those points were along the keyway that holds the cutter to the shank. From a design standpoint, the keyway is an abrupt change in geometry (i.e. a connection point) that has a direct impact on the systems flexibility.

Even though the abrupt change in dynamic flexibility along the Y-axis has no direct impact on the cutter performance in this case, it is important to recognize the ability to scientifically quantify such trends. It is possible to extend such readings to points on the machine and even on the workpiece. Such data would help isolate the actual source of an unstable cutting operation. A few other typical sources that would cause these trends are listed below:

- Faulty bearings in machine.
- Faulty spindle assembly.
- Uneven drawbar pressure.
- Improper fixturing of workpiece.

17.4.2 ELIMINATING CHATTER DURING CORNERING

A commonly overlooked source of chatter is machining of corners. As seen in ***Figure 145***, the engagement region is drastically increased from point "A" to "B". This in turn means that a greater number of teeth are in the cut, consequently changing the dynamic characteristics of the system.

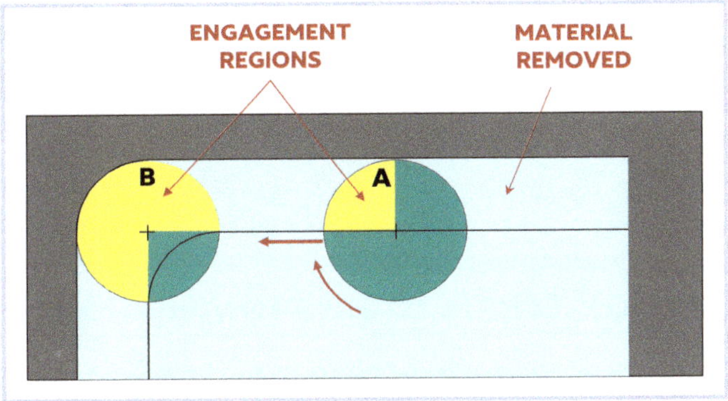

Figure 145. Engagement in Corners.

Such cornering instability problems may be solved my implementing an alternative tool path.

- Slowing down the Feedrate in the corners by as much at 50% will lighten the Chipload sufficiently to compensate for the additional teeth in the cut. Recognize that during cornering the cutter tends to act like it is in a Channel Cut. This can be undesirable from the standpoint that there may be rubbing at the start of the chip (especially in Positive Radial Rakes).
- Use spiral or half-moon shaped moves to rough out most of the pocket (***Figure 146***, Top). Then in the very end machine out a light periphery pass while slowing down the Feedrate in corners (***Figure 146***, Bottom).

- Kellering technique can also be used to widdle out the corners. The idea here is to excavate out a large corner by way of small arc like moves.

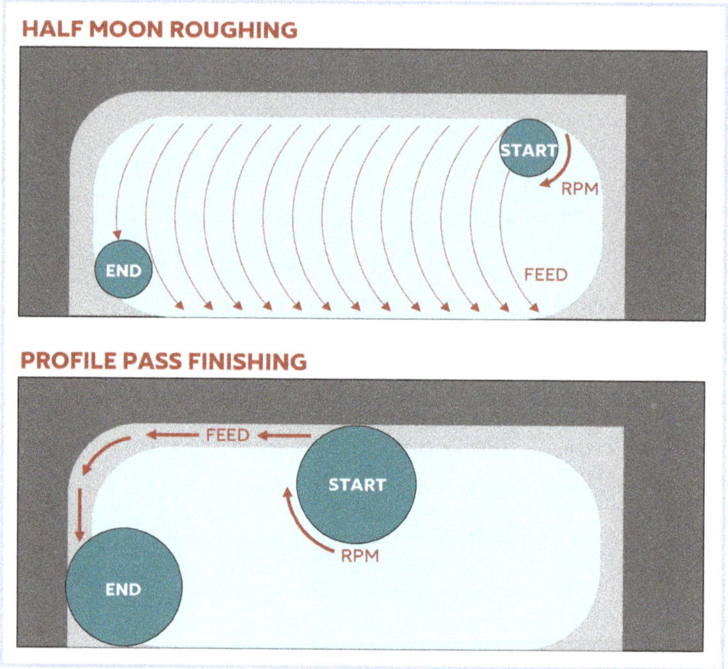

Figure 146. NC Strategy – Half Moon Roughing, & Profile Finishing Pass.

17.4.3 INCREASING FEEDRATE TO DAMPEN VIBRATIONS

This simple little trick of increasing Feedrate to dampen vibrations has been practices by many a machinist for numerous years. Let's say that during a milling operation Chatter kicked up, and there were no instruments available to measure it, nor was there available any expert with background knowledge on chatter theory. Here machinists have employed their years of cumulative knowledge & experience to resolving the Chatter by increasing the Feedrate. The natural tendency when an unfavorable chatter sound occurs, is to slow down. However, machinists know that often increasing the Feedrate will quiet the operation down.

Let's try to understand what is actually happening in this case. By increasing the Feedrate, the user is actually increasing the Chipload. This additional mass of chip acts as a dampener against the Regenerative Waviness that puts the chip out of Phase. This dampening does not actually put the chip back in phase, but rather cushions the cut. This technique works quite effectively, but is not exactly a root-causal type solution.

CAUTION: It should be recognized that if the cutter were cutting over complex changing geometries or corners, a change in engagement could make the chatter kick back up (and now at a higher Feedrate). That said, this simple little trick or technique can be quite effective over parts that have a mostly consistent cutter engagement condition.

17.4.4 ELIMINATING CHATTER DURING MACHINING THIN WALLED PARTS

The problem with machining thin walled parts is that the dynamics generated at the cutting tool engagement region are translated to the most flexible regions, which are typically the thin walls, ***Figure 147***. A commonly used technique to avoid such a problem is to relieve the tool. This prevents the already machined thin wall from getting in contact with the cutting tool and getting re-cut.

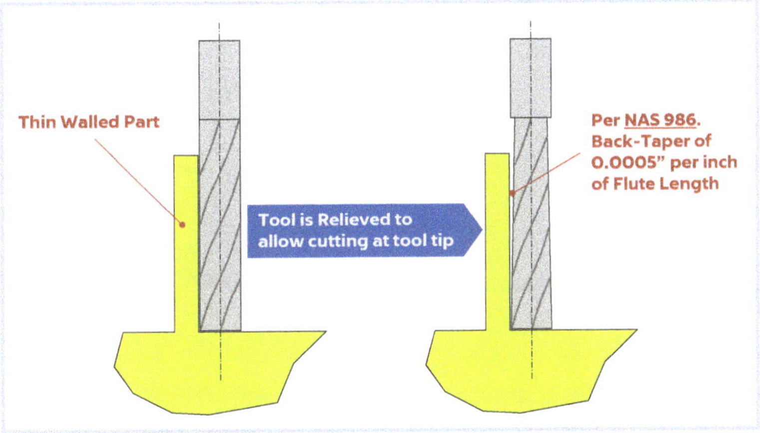

Figure 147. Machining Thin Walled Parts (Emphasis on Relieving Tool).

17.4.5 ELIMINATING CHATTER DURING MACHINING FLEXIBLE PARTS

Flexible parts as seen in ***Figure 148***, are a common source of dynamic instability. Some of the approaches to eliminate the instability arising from flexible parts are as follows:

- Program the tool path so as to maximize the local rigidity of the part. If possible, avoid the cutter path traveling over weakly supported areas to be machined. It is also a good practice to machine toward the fixture point.
- Machine the most flexible areas first, and the least flexible areas last.

- Avoid returning to an already machined flexible location.
- Shortening the tool usually stiffens all the modes of vibration. Therefore, use as short a tool as allowable.
- It is essential to use relieved tools when machining flexible parts. However, the amount of relief depends on the flexibility of the part.

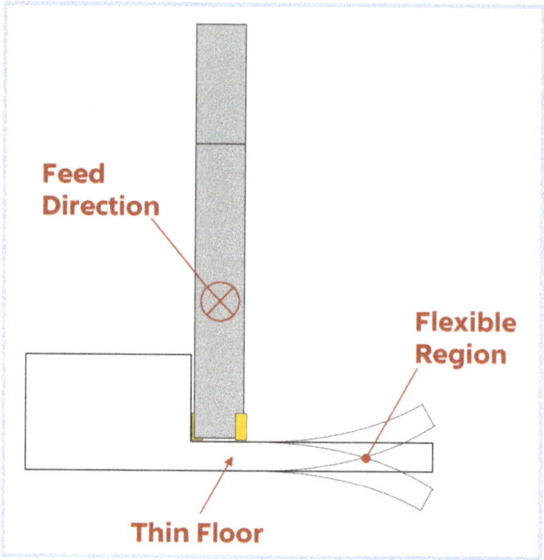

Figure 148. Machining Thin Floored Parts (Emphasis on Cutter-path Strategy).

17.4.6 ELIMINATING CHATTER AND IMPROVING CUTTING PERFORMANCE

- Shortening the overhang and tool usually stiffens all the modes of the system.
- Adding mass and dampers is also a way to enhance cutting performance. Such an approach is best when used on a system exhibiting a single flexible mode. Also, the damper should be positioned in the most effective area of the dynamic system.
- The cutting force distribution can be affected by shifting the forces from the flexible lateral direction, to the stiffer axial direction.

17.4.7 ELIMINATING CHATTER BY IMPLEMENTING SOME DESIGN GUIDELINES

- Since stability increases as tool length decreases, use the shortest tool possible.
- Stability increases as the tool diameter increases.

- Stability is not sensitive to spindle stiffness (diameter) unless the spindle becomes flexible and participates in the vibration mode.
- Stability is not sensitive to the tool holder length unless the holder becomes flexible and participates in the vibration mode.
- The spindle must be capable of operating at sufficient speed to achieve a stable operating point for the dynamically most flexible mode.
- High speeds require very tight control of tool balance.

Vibration phenomena are not random, but rather they can be quantitatively measured and described. With the aid of advanced technology, it is now possible to scientifically quantify the vibration characteristics of a milling process, predict chatter, and make recommendations to eliminate it. It is however important to recognize the degree of analysis that is required for a particular system. Chatter has a direct impact on surface finish, tool life, cycle time, and cost. It is important to understand the priority of one or more of these variables in a given operation is crucial to successfully eliminating chatter, and optimizing that specific milling operation.

18

TOOL WEAR

Tool wear initiates from the instant a cutting edge touches a workpiece. The study of tool wear is important from two fundamental viewpoint:

1. To recognize the type of tool wear.
2. To arrive at solution to either eliminate unfavorable wear, or to make the tool last longer.

The ultimate intention in learning about tool wear is to create cutting conditions that will yield repeatable-reliable tool wear in production.

18.1 TOOL WEAR BASICS

Before we dive into the study of the types of tool wear, there are some general tool wear basics that should the reviewed:

18.1.1 STUDY THE PROGRESSION OF WEAR

It is very important the user emphasize and study the progression of tool wear, rather than simply focus on the end of tool life. If we let a tool run long enough, its life will eventually end in chipping, or worst yet in catastrophic failure (tool breakage). The idea is to extend tool life by combating the initial failure mode, rather than the final one.

18.1.2 MEASURING TOOL WEAR

Figure 149 show the Rake and Flank faces of the cutter. As discussed in earlier chapter, the Rake Face is simply viewed as the face on which the chip would curl on. The Flank Faces are simply viewed as the clearance faces. During formal testing, the wear is always measured on the Flank Face. The wear should also be observed on the Race Face, and it is the user's discretion to measure on this face.

- Most failure modes will eventually result in chipping or fracture.
- The idea is to extend tool life by combating the initial failure mode rather than the final one.
- In formal testing, Wear is always measured on the Flank Face.
- Wear should also be observed on the Rake Face (measure optional).

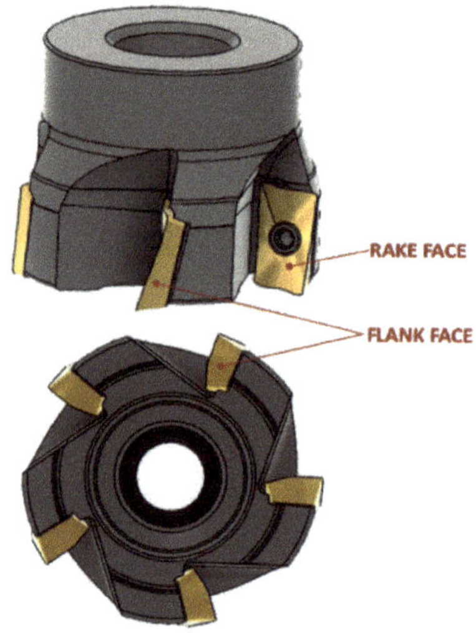

Figure 149. Tool Wear Basics.

18.2 TYPES OF TOOL WEAR

This section deep dives some of the most prevalent types of tool wear seen in the metal machining industry. Some of these wear types are preferred & favorable, while others should be avoided at all costs.

18.2.1 ABRASIVE WEAR

DESCRIPTION: Abrasive wear is one of the most preferred types of tool wear, ***Figure 150***. It the attrition of the primary clearance surface of an insert. It is caused by the rubbing action between the tool and workpiece. As seen in ***Figure 150***, Abrasive wear is very uniform in its appearance. Thus, making it very easy to analyze its progression, and measure its magnitude.

IMPROVEMENT TO CUTTER: Abrasive wear relates to the hardness characteristics of the cutting tool material. Abrasive wear can be improved upon by selecting a harder substrate material of the cutting tool. This can be achieved by either decreasing the Cobalt content, and or increasing the Tungsten Carbide (or alloys) content of the substrate.

- Attrition of the primary clearance surface of an insert.
- Caused by the rubbing action between the tool & workpiece.

- This is generally a preferred failure mode as it is predictable.
- Easy to measure and monitor its progression before catastrophic failure.

- Relates to Hardness of the cutting tool and workpiece material.
- Relates to SFM.

- Combat by selecting a harder or more wear resistant grade.
- Combat by reducing the SFM.

Figure 150. Tool Failure Modes – Abrasive Wear.

IMPROVEMENT TO CUTTING PARAMETERS: Abrasive wear relates to the heat signature and consequently the Surface Speed SFM of the cutting tool. Abrasive wear can be improved upon by reducing the SFM.

18.2.2 CHIPPING

DESCRIPTION: As the name implies, and shown in ***Figure 151***, in this failure mode a piece of the cutting edge is chipped off. In this text, Chipping is intentionally discussed early during the study of the types of tool wear. This is because just about all the types of tools wears eventually terminate in chipping. Chipping can leave undesirable finish marks on the workpiece. Chipping is generally categorized as Major & Minor Chipping.

Minor chipping can be viewed as a small nick to the cutting edge. If the user were to run their fingernail along a Minor chipped cutting edge, they may feel a small snag. In roughing operations, it is reasonable to continue cutting with minor chipping. Minor chipping usually measures less than about 0.005". Minor chipping may often accompany other types of tool wear as they progress. If a cutting insert is indexed (rotated) with minor chipping, the worn edge would still seat confidently in the cutter body's insert pockets.

Major Chipping is simply a catastrophic chip of the cutting edge. Here a significant porting of the carbide is missing. A Major Chip can measure from 0.015" or larger. If a cutting insert is indexed (rotated) with major chipping, the worn edge would still not seat confidently in the cutter body's insert pockets. This is because the edge boundary of a Major Chip could get deformed or raised with sharp peaks. In fact, if an edge with Major Chipping were indexed forcibly, it could cause gauging of the cutter body's insert pocket.

IMPROVEMENT TO CUTTER: Chipping relates to the toughness characteristics of the cutting tool material. Chipping can be improved upon by selecting a tougher substrate material of the cutting tool. This can be achieved by either increasing the Cobalt content, and or decreasing the Tungsten Carbide (or alloys) content of the substrate.

IMPROVEMENT TO CUTTING PARAMETERS: Chipping relates to the impact signature and consequently the Chipload APT of the cutting tool. Chipping can be improved upon by reducing the APT.

- Minor or Major chipping off of the edge material.
- Caused by impact issues or simply the terminal wear type of other wear patterns.
- Is generally categorized as minor & major chipping.
- Chipping can leave undesirable finish marks on the workpiece.
- Major chipping is usually the end of life.
- Relates to Toughness of the cutting tool and workpiece material.
- Relates to Chipload (APT).
- Combat by selecting a Tougher or more impact resistant grade.
- Combat by reducing the APT.

Figure 151. Tool Failure Modes – Chipping.

18.2.3 THERMAL CRACKING

DESCRIPTION: Thermal Cracking as seen in ***Figure 152***, is when cracks are formed on the rake or flank face of the cutting tool. These cracks are generally perpendicular to the cutting edge. Thermal cracks are cause by thermal expansion (when in cut), and contraction (when out of cut, & subject to coolant) of the cutting-edge during cutting (i.e. milling), ***Figure 153***. This phenomenon is known also as Thermal Shock.

Although, Thermal crack may appear unfavorable in appearance, their progression is actually very predictable. Thermal Cracks are also viewed as a preferred failure mode. The crack initiation and growth can be easily monitored. As small hairline cracks grow into more substantial one, additional hairline cracks start to form in between them. Ultimately the crack density is large enough to where the cutting-edge chips off between two dominant cracks.

IMPROVEMENT TO CUTTER: Marginal improvements in Thermal Cracking are to be had by changing the cutting tool material. If the progression of Thermal Cracking is unusually quick, the user should consider a more heat resistant cutting tool coating material.

IMPROVEMENT TO CUTTING PARAMETERS: Since Thermal Cracking is related to heat in the cut, it can be improved upon by reducing the Speed SFM.

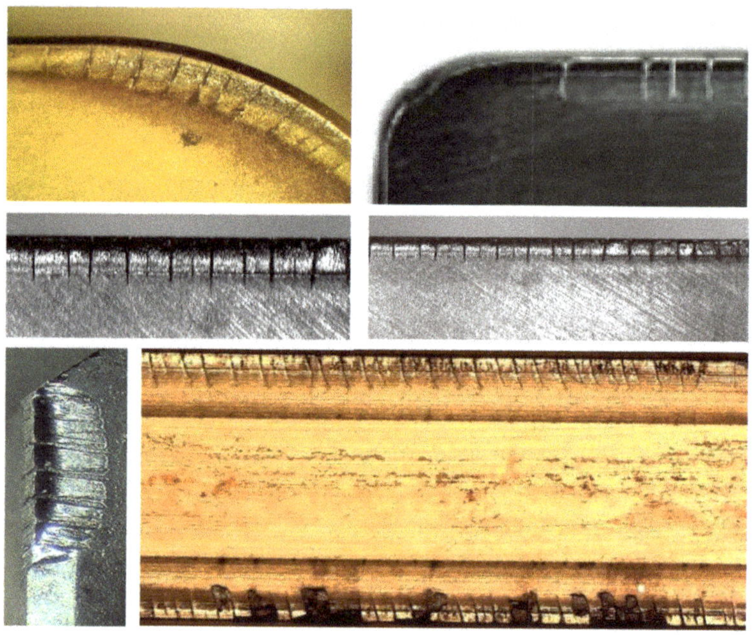

- Cracks formed on Rake or Flank face that are perpendicular to the cutting edge.
- Caused by thermal expansion & contraction during interrupted cutting (i.e. milling).

- This is a preferred failure mode, & its progression is predictable.
- Crack initiation and growth can be easily monitored.
- Eventually the material chips off between any two dominant cracks

- Relates to thermal shock to the cutting tool as it enters and leaves the cut.

- Combat by reducing speed SFM.
- Combat by turning coolant off, thus less gross temperature change.
- Combat by selecting coated carbide.
- Combat by selecting grade with higher thermal shock resistance.

Figure 152. Tool Failure Modes – Thermal Cracking.

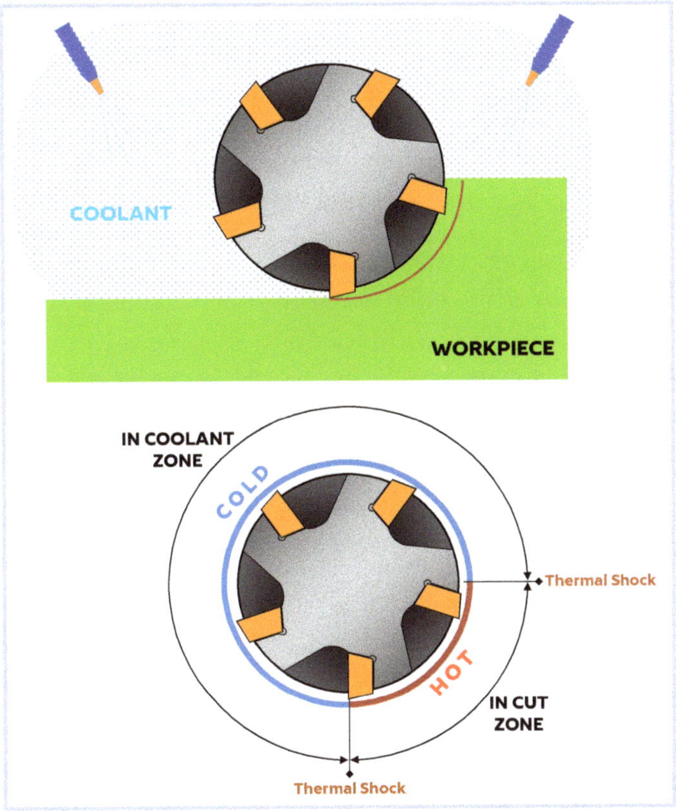

Figure 153. Understanding Thermal Cracking.

18.2.4 DEPTH OF CUT NOTCHING

DESCRIPTION: As the name suggests, Depth of Cut Notching, is a very localized notch (or chip) that occurs specifically at the Depth of Cut DOC, ***Figure 154***. Specifically visualize the transition point on the cutting edge, one side of which is immersed in workpiece material, and the other side is exposed to coolant, ***Figure 155***. This in turn creates a Thermal Shock from the temperature variation. The rest of the cutting edge would typically show abrasive wear. Depth of Cut Notching is not a preferred failure mode, but its progression is predictable. Often times, machine shops simply learn to live with it.

IMPROVEMENT TO CUTTER: Depth of Cut Notching can be improved upon by selecting a cutter grade with higher mechanical shock resistance. Increasing the edge preparation (hone size) can also improve reduce this failure mode. Since the Depth of Cut Notch is pretty much a type of chipping, occasionally selecting a slightly tougher grade could improve the magnitude of the notch. The **Caution**

in this case is that a tougher grade would inherently be less hard. Therefore, this could compromise the Abrasive wear on the rest of the cutting edge.

IMPROVEMENT TO CUTTING PARAMETERS: Since Depth of Cut Notch is a mode of Chipping, it relates to the impact characteristics of the cut. Therefore, the Notch progression can be slowed down by reducing the Chipload APT.

Another neat trick to reducing the magnitude of Depth of Cut Notching is to constantly move the location of the point where the notch occurs by constantly varying the Depth of Cut. This can be accomplished by feeding the cutting tool at a very gentle ramp angle or Sin Wave over the entire length of the cut. By constantly moving the location of the DOC, we are not allowing a singular point get stressed by Thermal & Mechanical Shock. Ultimately it is up to the NC programmer to determine be feasibility of such a cutter path strategy. Factors such a feature complexity, roughing or finishing, and workpiece stock amount have to be taken into account.

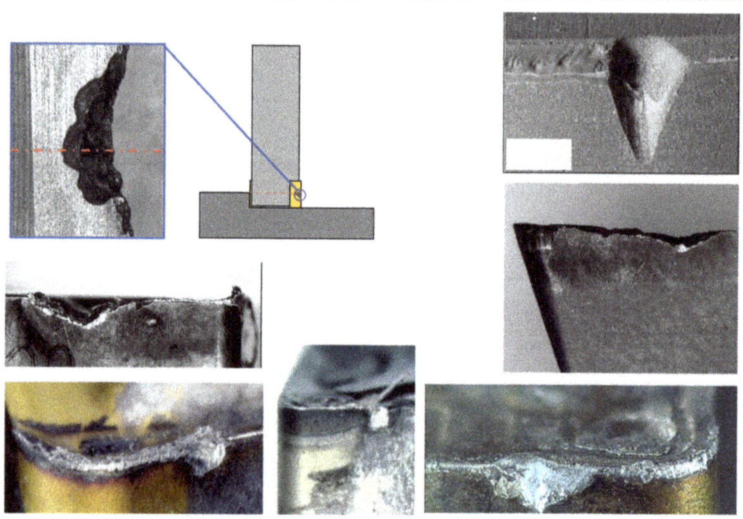

- Localized higher wear at the DOC line.
- Caused by higher surface hardness of the workpiece material.
- Caused by the thermal shock to the tool material where it is exposed out of the workpiece to either air or coolant.

- This is not a preferred failure mode, but its progression is predictable.
- Often times the shop learns to live with it.

- Relates to thermal and mechanical shock to the cutting tool at the transition point at the engagement in the workpiece and air (or coolant). This is typically the location of the DOC line.
- Relates to heat in the cut and toughness.

- Combat by selecting a grade with higher mechanical shock resistance.
- Combat by varying the Depth of Cut.
- Combat by increasing hone size or change geometry.

Figure 154. Tool Failure Modes – Depth of Cut Notching.

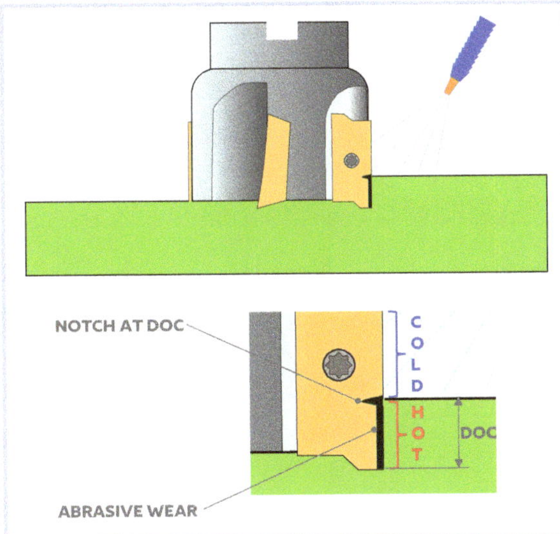

Figure 155. Understanding Depth of Cut Notching.

18.2.5 BUILT UP EDGE (BUE)

DESCRIPTION: As the name implies, Built Up Edge BUE, is a type of wear where the edge grows onto itself, ***Figure 156***. BUE is typically seen in soft metals machining (vis. Aluminum). During machining, temperatures in the cut are hot enough to where part of the chip gets molten. Often the parts that get molten are slivers of chips or chip particulates, rather than the entire chips itself. When the cutting edge exists the cut, and it exposed to coolant, the molten pieces of chip rapidly cool and weld on to the cutting edge. Thus, the cutting edge will actually grow onto itself. This has the appearance of a bound, or jagged cliffs on the cutting edge.

As the cutting continues, molten particulates continue to weld onto cutting edge and eventually create a condition where to tool is dimensionally altered (i.e. diameter). This could even affect the forces in the cut. Typically, the cutting edge grows a fixed amount and then naturally flakes, and the buildup process starts all over again. Machine shops get quite accustomed to this wear behavior, especially if it is cyclical and repeatable. BUE is certainly not a preferred failure mode, nor is it completely undesirable. It is simply a failure mode that machine shops tend to live with.

Occasionally, the built-up material does not conveniently flake off, and tends to be larger than average. Ultimately, when a larger than average BUE breaks off it tends to take a piece of the coating and carbide substrate with it. Now essentially the cutting tool itself is detrimentally compromised, can adversely affect the machining process.

IMPROVEMENT TO CUTTER: Considering the nature of this unique failure mode, there are marginal improvements that can be made to the cutting tool. The big idea here is to reduce the heat in the cut. Lighter edge preparations tend to be freer cutting, and could consequently create less heat. Additionally, high polished cutting tools, and slick coatings will tend to not let the chip stick to the cutter. Consequently, the chip particulates may have a lesser chance of welding to the edge.

IMPROVEMENT TO CUTTING PARAMETERS: The user should focus on parameters that reduce the heat in the cut. Reducing the SFM and APT would reduce the heat in the cut, and consequently reduce the molten condition of the chip. Visualize the arc of engagement during chip formation, and efforts that could shorten it. Factors such as WOC, and its proportion to Diameter should also be considered.

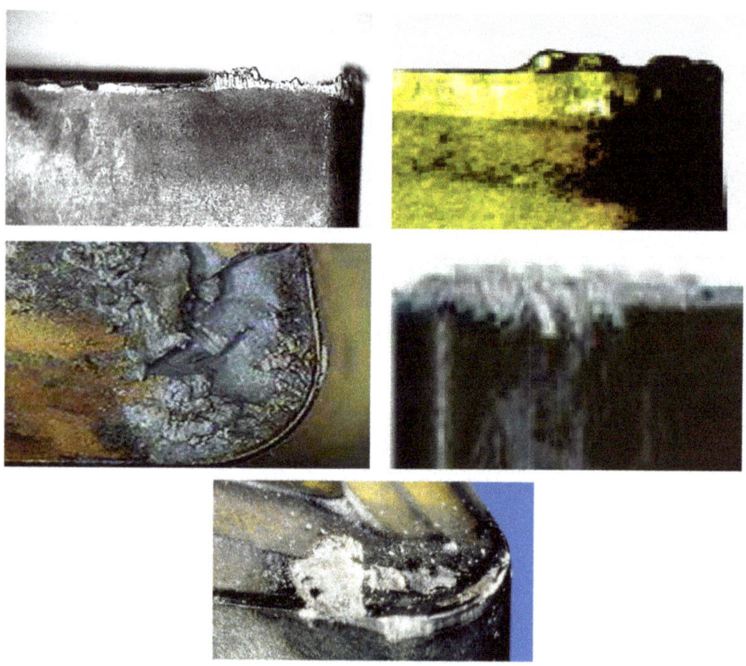

Figure 156. Tool Failure Modes – Built Up Edge (BUE).

18.2.6 DEFORMATION

DESCRIPTION: Deformation is a failure mode where the cutting tool material plastically deforms on account of the cutting forces ***Figure 157***. When the cutting edge deforms, it alters the cutting tool dimensionally, which in turn could be detrimental to the machining operation. Deformation is a completely undesirable failure mode, and the type of wear that you do not want to see. It is a sign that very poor choices were made in the selection & operation of the cutting.

IMPROVEMENT TO CUTTER: Deformation type wear is a sign that very poor choices were made in the selection of the correct cutting tool for the given workpiece. If you experience Deformation, stop machining. It is recommended to call upon a cutting tool company's service person, or technical expert to get a better recommendation. This recommendation will usually come in the way of a different carbide substrate & coating.

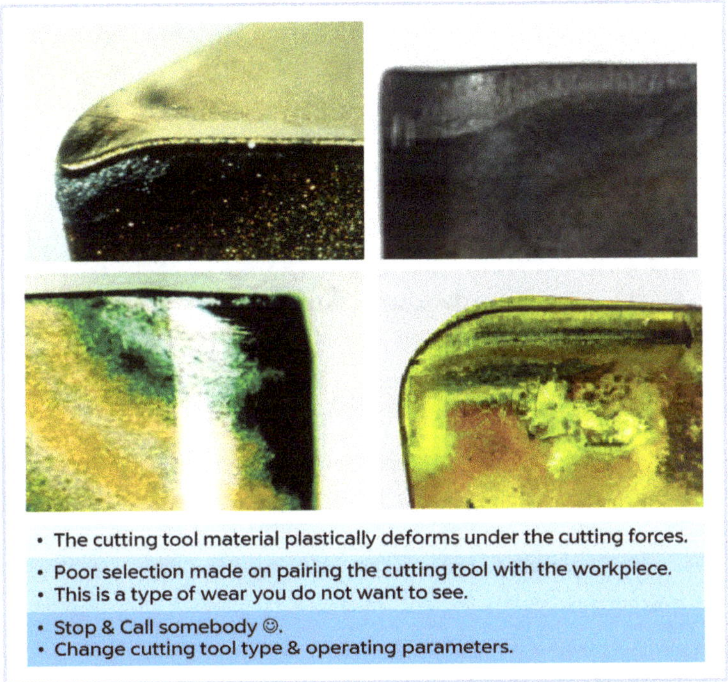

Figure 157. Tool Failure Modes – Deformation.

18.2.7 CHEMICAL / CRATER WEAR

DESCRIPTION: Simply put, Chemical / Crater wear is the interaction between the hot chip & the cutting tool rake face, ***Figure 158***. As the chip flows across the tool rake face, the carbide material could chemically dissolve into it. This type of wear is called Chemical Wear. Often chemical wear will also for a crater in the carbide rake face. There are also occasions where simply the shear friction of the chip on the rake face can form a crater. This wear is known as Crater wear. This type of wear is undesirable, and is not a preferred failure mode.

IMPROVEMENT TO CUTTER: Since this type of wear relates to the chemical affinity between the cutting tool and the workpiece material, it can be combated by selecting a coated carbide grade, and one that is more chemically inert. Changing the cutter's rake face geometry to reduce the forces can also help.

IMPROVEMENT TO CUTTING PARAMETERS: Since this type of wear relates to heat in the cut, it can be combated by reducing the SFM and APT.

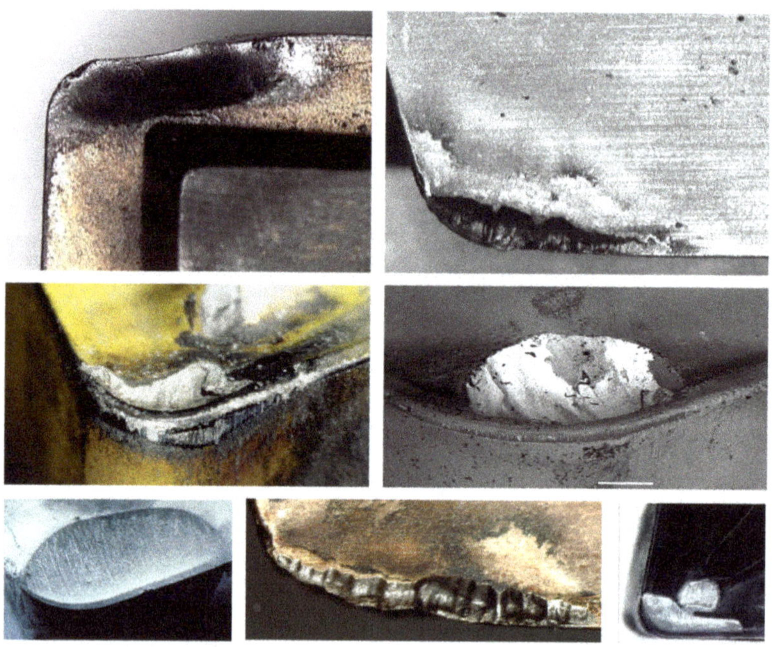

- As the chip flows across the tool's rake face, the Carbide material dissolves into it.
- Caused by a chemical interaction between the hot chip & the cutting tool.

- This is not a preferred failure mode.
- Severely impacts the machining process.

- Relates to chemical affinity between the cutting tool and workpiece material.
- Relates to heat in the cut.

- Combat by selecting a coated grade, & one that is more inert.
- Combat by reducing heat in cut (SFM, APT).
- Combat by changing insert rake face geometry to reduce forces.

Figure 158. Tool Failure Modes – Chemical Crater Wear.

The Structure of Diamond

- Each carbon atom is covalently bonded to four other carbon atoms.
- A lot of energy is needed to separate the atoms in diamond. This is because covalent bonds are strong.
- This also makes diamond's melting point and boiling point very high.
- There are no free electrons or ions in diamond, so it does not conduct electricity.

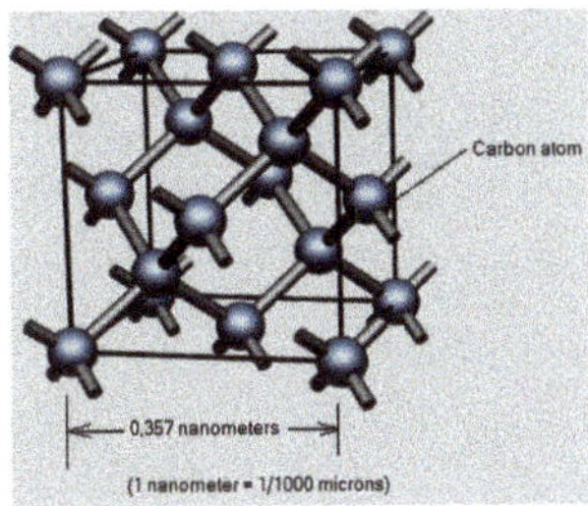

Poly Crystalline Diamond (PCD) Tools

- PCD tools are made of diamond, and diamond is essentially pure carbon, with atoms arranged in a cubic crystal arrangement that makes it very strong.
- At higher temperatures (>700 deg. C), the Carbon atoms react with Iron to form Iron Carbide (Fe3C).
  - They basically divorce their old Carbon partners in the crystal structure, get hitched to some new Iron partners, and the crystal structure collapses.
- With this sad exception of steel (and occasionally Titanium at high cutting temps), PCD does not bond with work piece materials. Because of this the built-up edge too is minimal.

Figure 159. Do Not Use Diamond Tools in Steel.

18.3 MEASURING WEAR DURING TESTING

As mentioned earlier, during formal testing tool wear should always be measured on the Flank Face of the cutter. The Rake Face should always be observed, and it is up to the user's discretion to measure here. Some typical wear analysis options are:

- Measuring the progression of wear on a single edge.
- Comparing the progression of wear affect by changing cutting parameters.
- Comparing the progression of wear between two or more cutters.

When it comes to measuring tool wear during testing, the most important rule is 'consistency'. Typically, a user is measuring a wear by way of a stereo microscope with a digital imaging system. Wear is typically measured from the outmost edge of the cutter, to the location of the wear line.

NOTE: Most of the time the user is interested in relative change in wear (i.e. Percentage change). We are always comparing wear to some reference value, so as to determine if we improved or hurt the process. From the viewpoint the more consistent the approach in the features selected for wear measurement, the more accurate the relative percent change.

18.3.1 SELECTING THE OUTERMOST EDGE OF CUTTER

As seen in ***Figure 160***, shows a classic abrasive wear pattern. At these high magnifications there can be some confusion or debate as to where to start measuring from. The features, shadows, and or discolorations in the cutting-edge region can be quite misleading. Some of these shadows are from the edge-preparation on the cutting edge (i.e. Hone, Lands or combos), as they move away from the focal zone of the microscope. This is where the user should recognize that ***measuring the wear is an art-form that yields scientific data***. With the intention to establish consistency of measurement this text recommends that the user always select the outermost (or front most) identifiable line of the cutting edge, ***Figure 160***. In doing so, we eliminate any guesswork in the start point of the wear measurement, especially if multiple measurements need to be taken as wear progresses. The next step in completing the wear measurement is to identify the wear line.

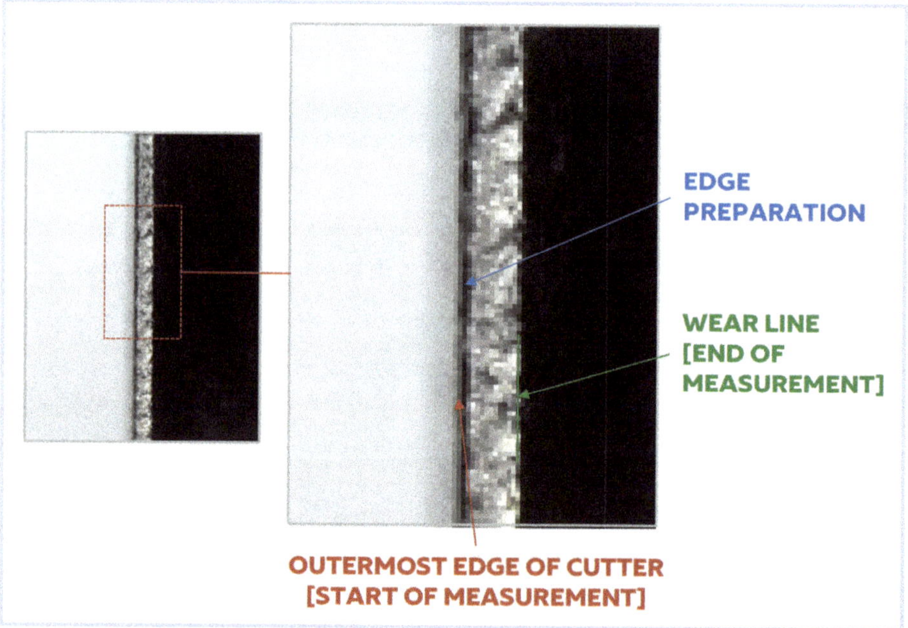

Figure 160. Measuring Wear – Selecting the Cutter's Edge, & Wear Line.

18.3.2 SELECTING THE WEAR LINE

Identifying a consistent wear line is key toward gathering value added wear data during testing. As seen in ***Figure 161***, the classic abrasive wear inherently carries some amount of texture and minor inconsistencies along its wear line. Although this text recommends some guidelines towards identifying a wear line, it is ultimately up to the user to decide on a wear location best relevant to their analysis. Some common wear line locations are:

- The location of maximum wear.
- The location of average wear.
- The location of wear where it makes the most consistent straight line.

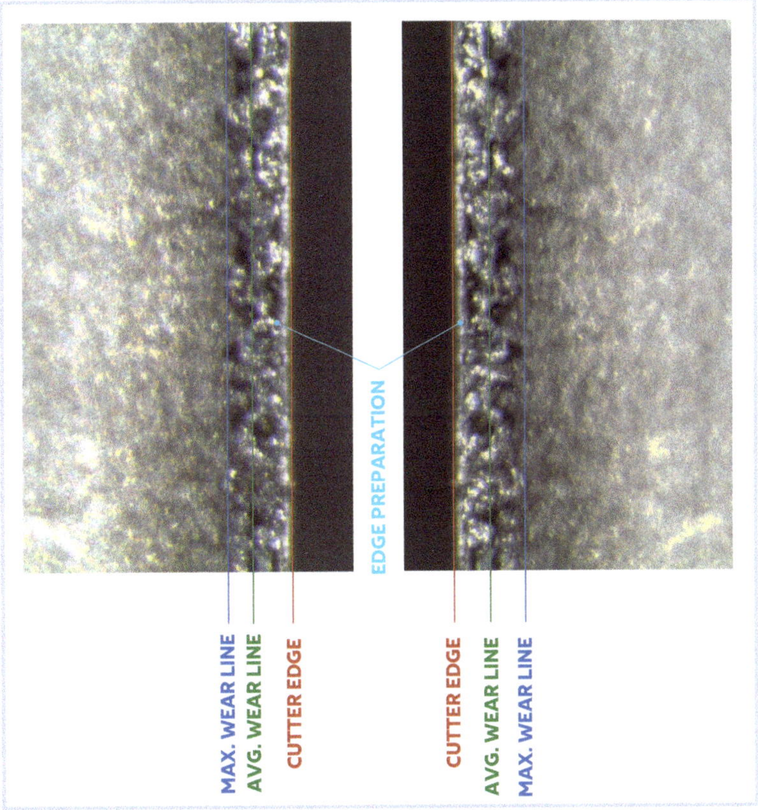

Figure 161. Measuring Wear – Average, & Maximum Wear Lines.

Often in most typical formal testing, the wear line is identified as the location makes the most consistent straight line. Additionally, it is also recommended that the user track any outstanding locations of maximum wear. This becomes particularly helpful as maximum wear could turn into chipping.

18.3.3 TYPICAL LOCATIONS ON CUTTING EDGE FOR WEAR MEASUREMENTS

To start with, the user is free to take as many wear measurements, at as many locations on a cutting edge as they deem necessary. This section simply lends experienced based input on typical location on cutting edges that become common talking points of wear analysis:

Solid Carbide Endmills

- Tangency of corner radius and the cutting edge.
- About a 1" zone past the region of tangency of corner radius mention above.
- Location of Depth of Cut DOC.
- Scan the cutting edge for any other abnormal wear.

Inserts on Indexable Cutters

- The lead insert (i.e. End station).
- The second insert (i.e. the insert directly behind the lead insert).
- The Insert at the Depth of Cut DOC location.
- Scan the entire flute for any other abnormal wear.

Drills

- The corners at the diameter.
- The central chisel point.
- Scan the cutting edge for any other abnormal wear.

Lathe Tools

- The corner radius of cutting point.
- Rake face along chip breaker geometry (for BUE, cratering, or chemical reactions).
- Flank face for measurements (or any unusual BUE).

19

CUTTING TOOL QUALITY TESTING

To complete the cutter manufacturing story, it is important to recognize some of the important quality checks that are routinely performed to ensure repeatable reliable performance of the cutting tool. The quality checks are partitioned into those for substrate and others for coatings. Any experienced machine operator will tell you that, ***'If something about a cutter does not look or feel right, it probably isn't right'***. Many years of troubleshooting machining operations, supporting production failures, testing processes, and lab analysis; have taught me that this simple intuition often line up with the sciences of analysis.

It's a mechanical world, and things will fail. Especially in the machining world, cutting tools are often are forefront of blame hierarchy. Often times cutting tools are not only the cheapest component in a machining setup, but also the most convenient to replace. Failures in machining processes may or may not be directly attribute to the quality of the cutting tool. Whether you are a user or a manufacturer of cutting tool, having a high-level recognition of some the quality check can support you in asking for the right data. Many a cutter warranty claim have been justified, or avoided with data from cutter quality check. It is not always necessary that a user of cutting tool have the related lab equipment to gather the data. These resources are conveniently available from university metallurgy labs, tech center, and cutter manufacturers. Other times, the user can conduct some basic checks with simple equipment such as a densitometer (density readings), and or a basic microscope.

19.1 SUBSTRATE QUALITY TESTING

Recognizing that the primary function of a substrate is to provide the hardness and toughness characteristics of the cutting tool, it quality and integrity becomes of paramount importance. The ability to manufacture thousands of cutting tools, while maintaining substrate consistency is a manufacturing marvel. It is one of the key contributors to achieving repeatable, reliable machining performance.

19.1.1 SUBSTRATE CARBON LEVELS

Carbon level is a measure of Tungsten diffused into the Cobalt binder material. Carbon levels in carbide materials contributes to both strength, and toughness for any given composition. When grades have either low or high carbon, they are almost totally useless for metal cutting. Considerable performance improvements (100+%) are achievable by targeting the carbon levels for any given composition and grain size. Nominal carbon level (stoichiometric) grades have high crack propagation resistance and decreased wear resistance. Sub-nominal carbon level (sub-stoichiometric) grades have high crack initiation resistance and increased wear resistance.

19.1.2 SUBSTRATE DENSITY

A density check is one of the first, quickest, & most common check that is done when a Carbide quality is suspected. Density of a material is defined as mass per unit volume. Density is a good confirmation means to ensure both macro porosity (a or b type) and composition are within allowable ranges. Density specifications are determined from exact metallurgical composition values with zero macro-porosity, and only 3% tolerance. Macro-porosity are pores over 25 microns (0.001") in diameter.

Density can be conveniently measured by an instrument called a Densitometer. It is basically a precision weighing scale which is capable of measuring weight of on an object in air and when suspended in water. Density is calculated as:

$$Density = \frac{Weight\ in\ Air}{(Weight\ in\ Air - Weight\ in\ Water)} \qquad \left[Units:\ g/cm^3\right]$$

19.1.3 SUBSTRATE POROSITY

Porosity is a measurement of microscopic voids within the material. Porosity is a manufacturing flaw within the material. As the size and amount of these flaws increase, strength & shock resistance decrease dramatically. Totally void free material is generally unachievable. ISO 4505 calls out the porosity ratings for an A and B spec carbide, ***Figure 162***.

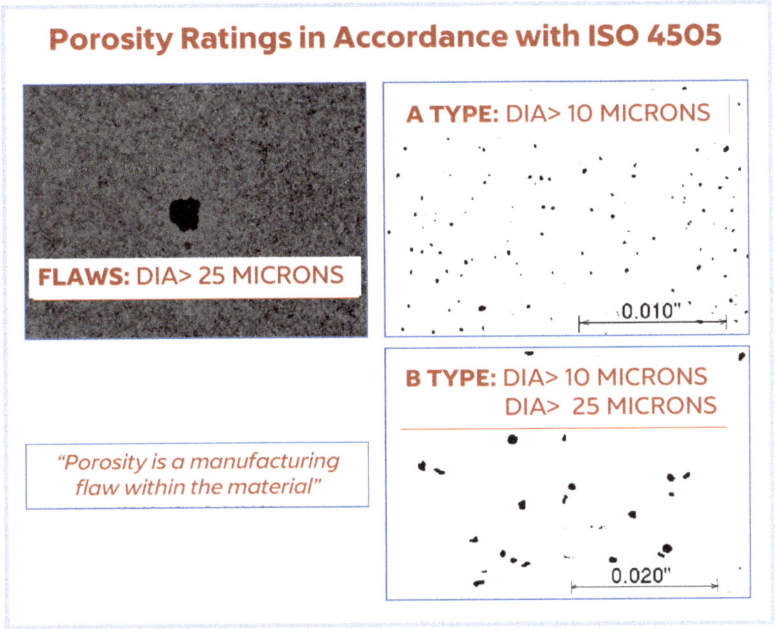

Figure 162. Porosity Rating – ISO 4505.

19.2 COATING QUALITY TESTING

This section talks about key tests that are used not only to evaluate the quality of the coating itself, but also the level of adhesion between the coating and substrate.

19.2.1 COATING ROCKWELL PUNCH TEST

Tests adhesion of PVD type coatings. A Rockwell "A" braille is used at 60 kg load. The punch forms a crater shaped indent on the surface. Poorly adhered coating peels off. As shown in ***Figure 163*** shows the insert that Passed the Punch test, as compared to the one that Failed.

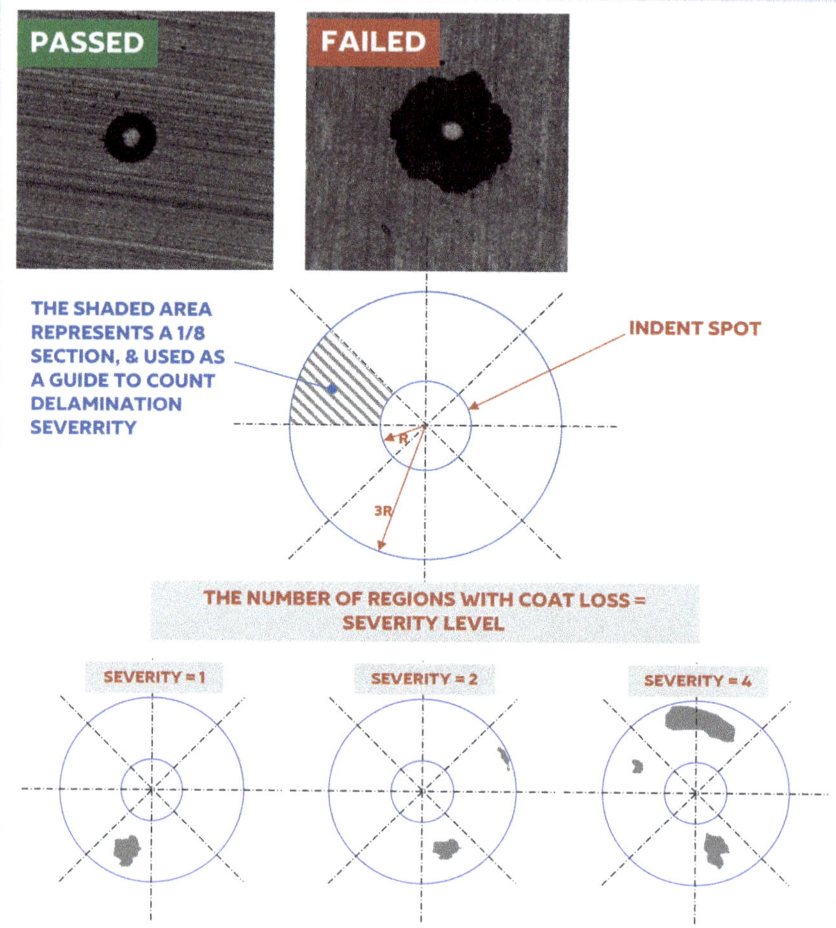

- Average severity of peeling must be less than '1'.
- No more than 6% of the sample may have a severity of 4 or higher.

Figure 163. Coating Quality Test – The Punch Test.

19.2.2 COATING BETASCOPE TEST

Tests thickness of any Ti(C,N) type coating. It could also be used for other single-metallic coatings (eg., CrC, possibly TiAlN, If it is homogenous).

19.2.3 COATING CALOT TEST

The Calot test checks for thickness, adhesion, and can check composition of any coating (an actual visual means of seeing a coating), ***Figure 164***. It is also useful in checking leach condition of the substrate. At higher magnifications, the Calot

Test shows coating adhesion as well as coating layer composition. It is also a geometric test of coating thickness. This test is also used to determine the extent of leaching, wherein cobalt is dissolved from the insert surfaces.

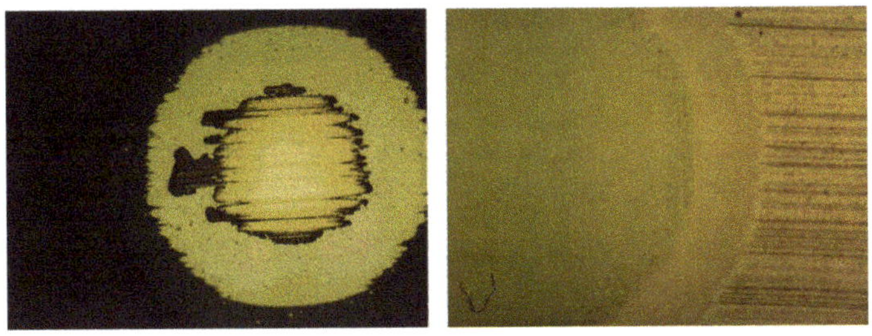

Figure 164. Coating Quality Test - Calot Test.

19.2.4 COATING RUB TEST

The rub test checks for adhesion of oxide coatings. Here the insert cutting edge worn by a diamond impregnated rubber wheel. This mode of coating removal shows the quality of the coating's adhesion. Abrasive wear is acceptable, while peeling is unacceptable. As shown in ***Figure 165***, the Top insert shows no peeling bottom insert shows consistent peeling of the oxide layer.

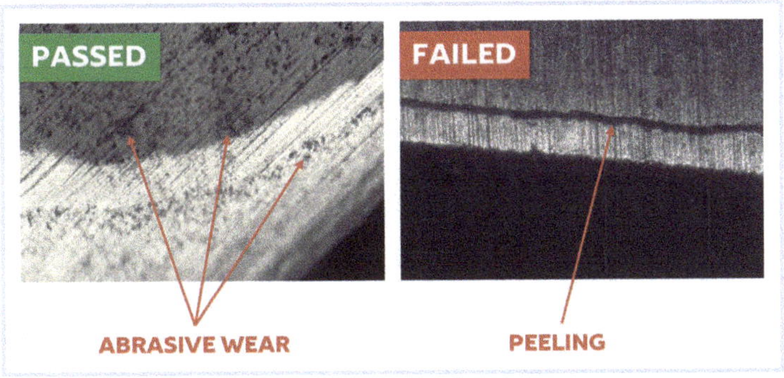

Figure 165. Coating Quality Test - The Rub Test.

19.2.5 LEACHING TEST

Leaching is the loss of cobalt from the surface of an insert. As shown in ***Figure 166***, is a side view of extremely severe (>10 microns) leaching. Leaching of 2 microns can cause coat failure.

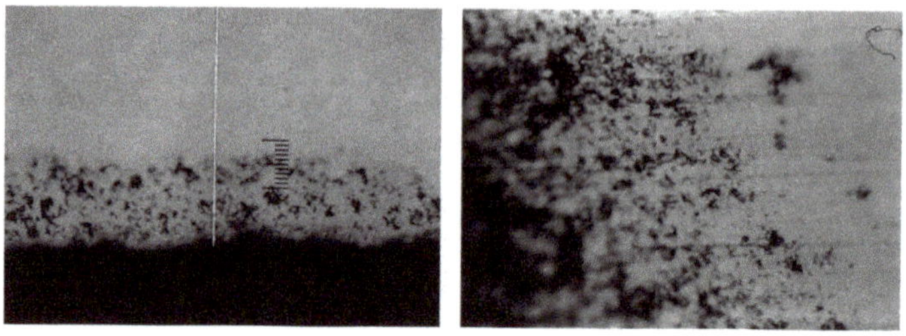

Figure 166. Coating Quality Test – Leaching Test.

19.2.6 COBALT CAPPING TEST

Cobalt Capping is when solid cobalt collects on the surface of an insert, ***Figure 167***. The Cobalt is drawn to surface during the sintering process. If the cutting tool is uncoated, there is not much detrimental effects. However, if the cutting tool is coated, Cobalt Capping can lead to the coating peeling off early in it usage.

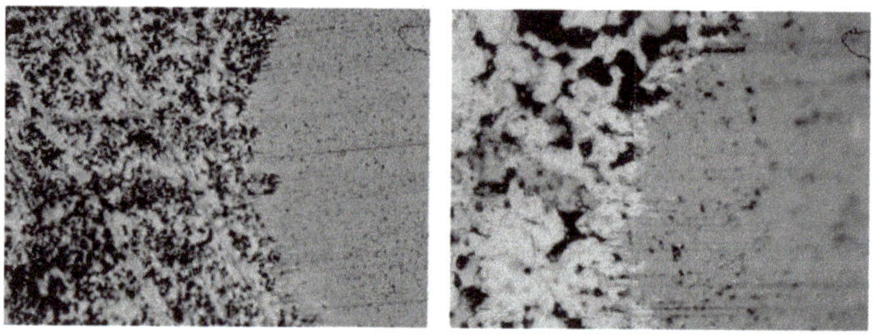

Figure 167. Coating Quality Test – Cobalt Capping Test.

19.2.7 ETA PHASE TEST

The Eta Phase is when there is a Carbon deficiency in the Tungsten Carbide (WC) and forms W2C. As seen in ***Figure 168***, it appears as black spots just beneath the coating layer. It is formed by insufficient Carbon in the substrate powder. One of the causes of this effect is improper atmosphere during the sintering process. This leads to decarburization in the coating.

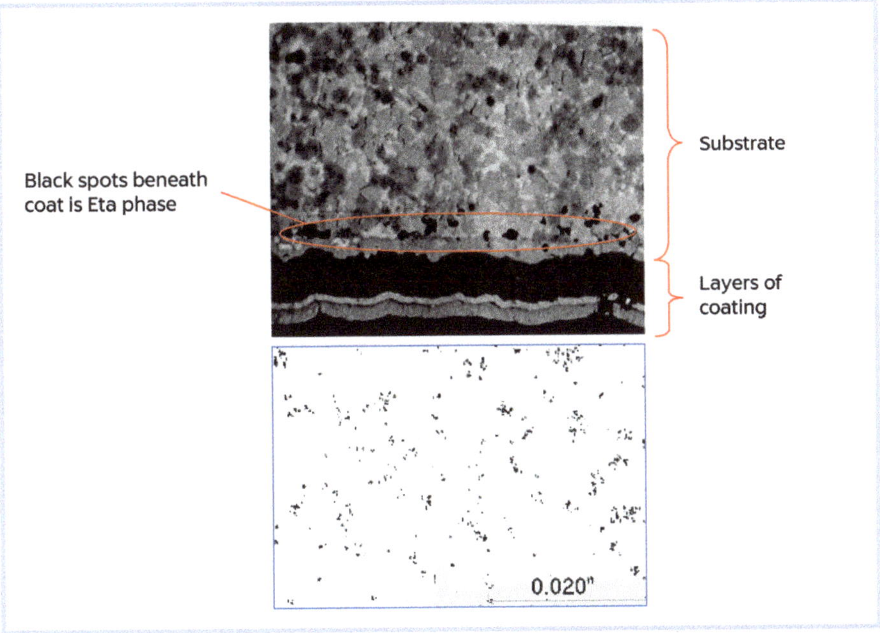

Figure 168. Coating Quality Test – Eta Phase Test.

20

FUTURE STATE OF MACHINING

The Art of Metal-Cutting has truly survived the test of time. From the era of early blacksmithing, to modern day CNC machining, it has evolved through the necessity of Industrial growth and revolutions. The demands of numerous wars have been one of the key contributors to significant technological growth in machining. Machining touches just about everything that makes our lives possible (homes, transportation, agriculture, medical, entertainment, sports, and countless other fields). With hopes of continuous improvements, we are simply trying to do our very best to attach pieces of science and math to this magnificent art form.

The flip side of the coin is that we live in a consumer-based, materialistic society. As we continue to build things, we also deplete the natural resources, and use up significant sources of energy. In recent years, there has been a heightened awareness and sensitivity to the availability of raw materials (both for workpiece, and tools that cut them). The manufacturing community is continuously evolving methods and techniques that need less materials, and use less energy. This chapter strives to get a discussion started on the evolution and future state of machining.

20.1 NEAR NET SHAPE PARTS

Near Net Shape simply implies that the starting condition of the workpiece material is as close to that of the desired finished part as possible. The closer the starting workpiece is to the finished part, the less material we have to remove, and therefore we utilize less energy / resources.

FOR EXAMPLE: As discussed in the section on Material Removal Rates, the ratio of the weight of the forging compared to the weight of the finished part is known as the Buy-to-Fly Ratio. In certain large parts required for automotive, aerospace, die & mold industry; this Buy-to-Fly ratio can be as large as (10:1 or higher). Which means that if the forging or block weight was say 4000 lb, then the weight of the finished part could be 400 lb. This works out removing 3600 lb of chips.

Simply halving the above buy-to-fly ratio down to 5:1, would mean the weight of the forging or block would be 2000 lb. Since this would still be machined down to 400 lb, the weight of the chips would now be 1600 lb. As we reduce the amount of chips that need to be removed, we not only reduce the footprint on the raw material starting condition, but also significantly cut down on the cycle time for machining.

The idea of Near Net Shape is to reduce the Buy-to-Fly ratio as much as possible. A dramatic scenario in the above example would be a ratio of 2:1, or less. Recognize that a Buy-to-Fly ratio of 1:1 simply means we have nothing to machine.

NOTE: One cannot help but wonder why in the current state, the weight of forgings is so much higher than that of the finished part. The justifiable answer is that: in order to achieve the material quality and consistency in the finished part, the forging simply needs to be heavy. In other word a quality finished part is buried deep inside a raw forging. This could be a sheer manufacturing limitation of the forging industry.

The unjustifiable/unacceptable answer to the above question is that: there is no reasonable business model for a forging manufacturer to reduce forging weight. Why should they? The heavier the forging, the greater the cost at which they can sell it. Simply put, a forging yard needs to be incentivized in order to reduce the Buy-to-Fly ratio. And no reasonable incentive has been proposed to date.

20.1.1 ADDITIVE MANUFACTURING

Just as machining is a type of subtractive manufacturing (where material is cut out to create the finished part), additive manufacturing is where the material is added on to create the finished part. There are various forms of additive manufacturing that will be discussed briefly in the next subsections.

20.1.1.1 RAPID PROTOTYPING

Rapid prototyping is a family of innovative additive manufacturing methods that is used to build models of 3D parts. Some of these models can even be functional parts. The most common of these techniques is termed 3D printing. Starting with a CAD model, generally rapid prototyping works on the principles of additive manufacturing by layers. Some popular rapid-prototyping techniques are listed as follows:

- 3D printing (3DP).
- Ballistic particle manufacturing (BPM).
- Directed light fabrication (DLF).
- Direct-shell production casting (DSPC).
- Fused deposition modeling (FDM).
- Laminated object manufacturing (LOM).
- Laminated resin printing (LRP).
- Shape deposition manufacturing (SDM) (and Mold SDM).
- Solid ground curing (SGC).
- Selective laser sintering (SLS).
- Selective laser melting (SLM).
- Stereo lithography (SLA).
- Multi Jet Fusion (MJF).

20.2 GREEN INITIATIVES

Sensitivity to raw material availability, power consumption, and energy efficiency are significantly heightened. Raw materials are not limitlessly abundant. The simple fact is that all material things came from our planet, and at some point, gluttonous consumption will lead to their depletion. More than ever, the need for recycling and repurposing of manufactured products is of paramount importance. To say that 'we as a human race will champion manufacturing of future products, but not care about the starting condition of the amount of material needed', is simply as irresponsible as 'putting more food on your plate than you can or intend to eat'. This approach is irresponsible, and downright unacceptable. It is highly likely, if not even necessary, that we will continue to invent new manufactured materials with unique material properties. The machining community will simply have to adapt to this this change in tides.

Just as the availability of raw materials, the energy needed to manufacture these into final products is neither abundant nor cheap. The power consumption during manufacturing plays a significant role in the bottom line of many an industry. A simple adage is that a lighter machine component draws less energy to move around and manipulate. Analogously, lighter finished products draw upon less energy to function, and consequently produce less environmentally harmful emissions.

20.3 FUTURE STATE OF MACHINING

Although I do not profess to have any sort of crystal ball, the opinions expressed in the following subsections are mere educated hypothesis. These opinions are formed by observing recent trends, and studying upcoming industries. Attention is also given to industrial pain points, and sensitivities to material availability. These educated hypotheses are simply based on understanding the impact of one variable change in the inputs of machining, to those variables affecting the outputs of related manufacturing processes. This future state of machining is all interconnected (strategy, tools, machines, manufacturing) and shall be discussed as such, collectively:

Machining Strategy: Assuming that we get to a state of significantly reduced Buy-to-Fly ratios, machining strategies would have to evolve accordingly. With lesser material to machine, lesser is the need for roughing. As we approach Near-Net-Shape, the need for aggressive roughing would be greatly diminished if not eliminated. Machining strategy focus would shift to semi-finish and finishing needs only. At a very high level, machining engagement would turn lighter.

Shallow & Fast: Lighter cutting would mean that there would be a heightened trend towards shallow-and-fast machining. Imagine a future state where chips are no longer the shapes of 6s and 9s, but rather slivers (the kind that get under your skin and can irritate incessantly).

Speed: Shallow and fast cutting would lead to less heat in the cut, and consequently manufacturers would trend towards speeding up the cuts. Imagine a future state where the chosen Surface Speeds (SFM) and consequently RPMs are on a magnitude of ten times (or more) faster than those the current high velocity techniques.

Hardness Over Toughness: Significantly lesser aggressive roughing in the future state could yield lesser dependence on the toughness (ability to resist impact) properties of cutting tools, and more so on the hardness (ability to resist abrasion). Imagine this future state, with a drive for significantly harder cutting tool substrate materials, and how it would impact substrates and coatings.

Cutting Tool Materials: With the future need for significantly higher speeds, and harder tools, the cutting tool industry would have to evolve drastically. The seeds of this future evolutions are already in germination. This evolution can be seen in: the advent of new alloying materials, tighter tolerances to substrate material quality, extremely micro elements, binder-less carbide, and enhancements to coating adhesions. **Coating** would have to withstand significantly higher cutting temperatures.

Cutting Tool Geometries: The geometries in the futures state of cutting tools will not only have to accommodate the higher speeds and feeds, but also be able to confidently create a sliver type chip consistently. The edge-preparation would evolve in a way to support consistent chip starting condition, while precisely redistributing stresses and forces that occur during the future higher velocity machining.

Tool Holding: The need for significantly higher RPMs in close to finishing operations would drive out enhanced tool holding systems. They would be able to accommodate precision balancing with consistent repeatability. Confident tool gripping would be vital, and no tool pullout would be acceptable. In the future state the holding systems would be extremely symmetrical with vibration dampening characteristics.

Attention to Best Practices & Details: In the future state of precision high velocity machining, all the best practices would take on a heightened priority. Tools would have to be mandatorily precisely **Balanced**, possibly to an even tighter G value than currently (possible G = 1 or lower). **Gage Lengths** would have to be precisely maintained to an even tighter tolerance. **Vibration** sensitivity at the higher RPMs would be extremely heightened. At the higher RPM the need for **Advanced Dynamic Analysis** (Tap Testing) would become of paramount importance so as to avoid any vibrational contribution.

Machine Tools: In the future state cutting tools are going to need machine tools that can drive them at significantly higher velocities, and yet maintain the extremely tight tolerances required by the near net shape part. The machine **Controllers** would evolve such that they could efficiently keep up with processing the NC code. To move at such high speeds the machines would have to be lighter and nimble, while maintaining the desired stiffness and rigidity. Evolution in acceleration & deceleration requirements from machine feeds will be a significant leap forward for the **Axis Drive** technology. To accommodate this combination of characteristics, it is very likely that industry will start manufacturing just the right machine tool, for just the right specific part.

Hybrid Machines: Additionally, in the future state it is very likely that machine tool will be a blend to additive and subtractive technologies. After all, the NC codes required for both technologies are quite similar. It is very likely that machine tools would grow a Near Net Shape part, and then execute the finish machining of the final product. Keeping the part anchored in the same setup always allows for maintaining tighter tolerances. This would be of particular importance in creation of large structural parts.

Power vs Torque: The combination of the increasing speeds of the machine tools, and the heightened need for machining near net shape parts (shallow and fast), will create a greater dependency on the need for machine spindle Power rather than Torque. This need for focused power will drive the evolution of newer **Spindle** technologies.

20.4 THE FUTURE RELATIONSHIP OF ADDITIVE MANUFACTURING AND MACHINING

In recent years, there is a palpitating tension, and even an unspoken excitement amongst machinists, as the manufacturing industry makes its turn towards true near net shape parts. With each passing year, the advent of rapid prototyping is taking a firmer hold towards true production ready parts. A severe leap from its roots in simply prototype scaled models, visual aids, and toy pieces. Combined with more advanced additive processes like welding, carbon fiber reinforced polymers (CFRP), friction stir welding, and robotics; additive manufacturing is making imprint as the mode of future operations. Does this mean that machining is irrelevant? Does machining disappear? The answer is no, absolutely not. These two industries need not be looked upon as in two separate competitive silos. Rather, machining will become a vital complementary industry to support the future state of additive manufacturing. The machining trade is on the precipice of a significant technological revolutions to accommodate the demand for this change.

The future change in machining technology is exciting but also challenging. It will strain on the machine builders for significantly faster speeds and accuracy, challenge the cutter manufacturers with needs for dedicated materials, and lend heightened importance to observing all the best-practices in machining. This challenge is by no means and easy one. I for one, am truly looking forward to what's possible in our future.

GLOSSARY

TEXT	ISO	NAME	IMPERIAL	METRIC
APR	f_n	Advance Per Revolution	inch	mm
APT	f_z	Advance Per Tooth (Chip Thickness)	inch	mm
b_{Limit}	-	Limiting Depth of Cut	inch	mm
B_m	M_{max}	Bending Moment	lbf.ft	N.m
CAT	-	Caterpillar (Originated Holder).	-	-
C_L	-	Clearance Length	inch	mm
CTM	-	Chip Thinning Multiplier	-	-
DIA	DC	Diameter of Tool	inch	mm
D_{EFF}	DC_{ap}	Effective Diameter	inch	mm
DOC	a_p	Depth of Cut (Axial)	inch	mm
# EFF	z_c	Number of Effective Edges	-	-
F_L	-	Flute Length	inch	mm
FRF	-	Frequency Response Function	-	-
F_T	F_T	Total Cutting Force	lbf	N
F_U	-	Force due to Unbalance	lbf	N
f_t	-	Tooth Pass Frequency	Hz	Hz
G	G	G-Scale for Level of Precision of Acceptable Unbalance	-	mm/s
G_L	G_L	Gage Length	inch	mm
HP	P_c	Power	Hp	kW
HSK	-	German: Hohl Schaft Kegel, translates: Hollow Shaft Cone.	-	-
Hz	Hz	Frequency or Chatter Frequency	Hz	Hz
ID	-	Inside Diameter	inch	mm

TEXT	ISO	NAME	IMPERIAL	METRIC
Im[G]	-	Imaginary Part of Frequency Response Function	-	-
IPM	v_f	Inches Per Minute (Feedrate)	in/min	mm/min
K	k_c	Power Factor	Hp min/in^3	kW min/ m^3
k	k	Stiffness	lbs/inch	N/m
KG	-	Weight of Tool & Holder Assembly for Acceptable Unbalance	-	kg
L	-	Chatter Marks Spacing	in	mm
MRR	Q	Material Removal Rate	in^3/min	cm^3/min
N	-	Lobe Number	-	-
OAL	-	Overall Length of Tool & Holder Assembly	in	mm
OD	-	Outside Diameter	in	mm
R	-	Radius of Tool	in	mm
R_{EFF}	-	Effective Radius	in	mm
r	r	Ratio of Excitation Frequency to Natural Frequency	-	-
Re[G]	-	Real Part of Frequency Response Function	-	-
RPM	n	Revolutions Per Minute	rev/min	rev/min
SFM	v_c	Surface Feet Per Minute (Speed)	ft/min	m/min
TQ	M_c	Torque	lbf.ft	N.m
U	U	Acceptable Unbalance	-	g.mm
WOC	a_e	Width of Cut	in	mm
ζ	ζ	Dampening Ratio	-	-
π	π	Pi constant (3.1415)	-	-
μ	μ	Directional Orientation Factor	-	-
ω	ω	Excitation Frequency	Hz	Hz
$\omega\nu$	$\omega\nu$	Natural Frequency	Hz	Hz

INDEX

www.ingramcontent.com/pod-product-compliance
Lightning Source LLC
LaVergne TN
LVHW071340210926
848486LV00003B/200